Clustering and Information Retrieval

Network Theory and Applications
Volume 11

Clustering and Information Retrieval

Weili Wu

Department of Computer Science,
The University of Texas at Dallas,
Mail Station EC 31,
Box 830688,
Richardson, TX 75083, U.S.A.

Hui Xiong and Shashi Shekhar

Department of Computer Science and Engineering,
University of Minnesota - Twin Cities,
EECS BLDG 4-192,
200 Union Street SE,
Minneapolis, MN 55455, U.S.A.

Distributors for North, Central and South America:
Kluwer Academic Publishers
101 Philip Drive
Assinippi Park
Norwell, Massachusetts 02061 USA
Telephone (781) 871-6600
Fax (781) 871-6528
E-Mail <kluwer@wkap.com>

Distributors for all other countries:
Kluwer Academic Publishers Group
Post Office Box 322
3300 AH Dordrecht, THE NETHERLANDS
Telephone 31 78 6576 000
Fax 31 78 6576 474
E-Mail <orderdept@wkap.nl>

Electronic Services <http://www.wkap.nl>

Library of Congress Cataloging-in-Publication

Wu, Weili/ Xiong, Hui/ Shekhar, Shashi
Clustering and Information Retrieval
ISBN-13: 978-1-4613-7949-2 e-ISBN-13: 978-1-4613-0227-8
DOI: 10.1007/978-1-4613-0227-8

Printed on acid-free paper.

Contents

Foreword

Clustering is an important technique for discovering relatively dense sub-regions or sub-spaces of a multi-dimension data distribution. Clustering has been used in information retrieval for many different purposes, such as query expansion, document grouping, document indexing, and visualization of search results. In this book, we address issues of clustering algorithms, evaluation methodologies, applications, and architectures for information retrieval.

The first two chapters discuss clustering algorithms. The chapter from Baeza-Yates et al. describes a clustering method for a general metric space which is a common model of data relevant to information retrieval. The chapter by Guha, Rastogi, and Shim presents a survey as well as detailed discussion of two clustering algorithms: CURE and ROCK for numeric data and categorical data respectively.

Evaluation methodologies are addressed in the next two chapters. Ertöz et al. demonstrate the use of text retrieval benchmarks, such as TRECS, to evaluate clustering algorithms. He et al. provide objective measures of clustering quality in their chapter.

Applications of clustering methods to information retrieval is addressed in the next four chapters. Chu et al. and Noel et al. explore feature selection using word stems, phrases, and link associations for document clustering and indexing. Wen et al. and Sung et al. discuss applications of clustering to user queries and data cleansing.

Finally, we consider the problem of designing architectures for information retrieval. Crichton, Hughes, and Kelly elaborate on the development of a scientific data system architecture for information retrieval. Their approach is to build a system solution that allows for the clustering and retrieval of information to support scientific research. In the final chapter of the book, Yao presents the design of an information retrieval support systems (IRSS) using granular computing. IRSS is expected to be another general framework for supporting scientific research.

We wish to thank all those who contributed articles or reviewed articles for this book. We believe this collection of articles will serve as a useful reference in bridging the gap between clustering and information retrieval.

Weili Wu
Hui Xiong
Shashi Shekhar

CLUSTERING AND INFORMATION RETRIEVAL (pp. 1-33)
W. Wu, H. Xiong and S. Shekhar(Eds.)

Clustering in Metric Spaces with Applications to Information Retrieval

Ricardo Baeza-Yates
Benjamín Bustos
Center for Web Research, Dept. of Computer Science
Universidad de Chile, Blanco Encalada 2120, Santiago, Chile
E-mail: {rbaeza,bbustos}@dcc.uchile.cl

Edgar Chávez
Universidad Michoacana, Morelia, México
E-mail: elchavez@fismat.umich.mx

Norma Herrera
Univ. Nacional de San Luis, San Luis, Argentina
E-mail: nherrera@unsl.edu.ar

Gonzalo Navarro
Center for Web Research, Dept. of Computer Science
Universidad de Chile, Blanco Encalada 2120; Santiago, Chile
E-mail: gnavarro@dcc.uchile.cl

Contents

1 Introduction

The concept of cluster is elusive. The direct implication of this is a large diversity of clustering techniques, each serving a particular definition. Most techniques focus on a global optimization function. The general procedure is to propose a clustering (using a suitable algorithm), then to measure the quality and amount of clusters, and repeat the procedure (proposing a new clustering structure, using for example new parameters) until satisfied.

This setup is satisfactory for many applications. Traditional clustering has been used in Information Retrieval (IR) for many different purposes, such as query expansion, document categorization and classification, relevance feedback, visualization of results, and so on. However, there are a number of applications in IR where new clustering methods would be useful.

In this chapter we propose a non-traditional clustering method, which is aimed at clustering data from a general *metric space*. This means that all the information we can use for discovering the clustering structure is the matrix of distances between all the data pairs in the set. Moreover, we cannot create new data points such as centroids. This is in contrast with traditional clustering methods (such as the k-means technique), which operate on coordinate spaces and may require to create new data points.

As a result, our technique is more general and can be used in applications where traditional clustering methods cannot be applied.

Our approach is aimed at clustering multimedia data. For this we mean data with no coordinates or dimensional information whatsoever. Several clustering algorithms exist for this scenario, for example, graph theoretical methods such as Zhan's algorithm [Zha71]. However, in several IR applications we need to go further. First, we may need to obtain the cluster of a given element (which will be regarded as its "equivalence class") without having to compute the clusters for all the data. In applications such as stemming and conflation we just want to obtain the class of a given word, which does not depend on the surrounding data. Even more difficult, we may want to compute the class of a word not in the data set (think, for example, of a misspelled word). A second requirement is the need to compute a *cluster hierarchy* by scaling a parameter. This hierarchy must start with the individual elements and end (possibly) with an equivalence class of the entire sample.

A third and very important requirement is that of having a *robust* clustering method. Data samples are often contaminated with spurious data objects, that is, objects not following the same distribution as the rest. These are called *outliers.* Deciding that an object is an outlier is rather risky[1]. We require a robust method that assigns an equivalence class for each element. Querying the system with any object will produce its equivalence class, and the result must be the same for any object in the class used as a query. If the equivalence class of the object is the object itself, then we will call it an outlier.

In the next section we present our clustering technique for metric spaces. Then, we show three different applications of clustering in IR, all them under the metric space model.

In Section 3 we apply the clustering method to the problem of conflating words that share a common root, which is a relevant task in IR. The standard solution for stemming is a rule based approach inherited from Porter's algorithm. This type of solution closely follows the possible variations of a single stem into all its morphological instances. This technique suffers from several drawbacks, in particular its inability to cope with noisy data. For example, when a word is misspelled, finding its correct stem by following rules is at least risky. A single edit error in a word may trigger the wrong rule

[1] There are a number of horror stories about automatic data cleaning (outlier removal) that show that, under an unknown distribution, it is better to be cautious. An example is the ozone layer measurements, where the computer filtered out large holes as outliers, but the holes were actually there.

and cause a mistake in the conflation. We propose a more robust solution based in clustering on metric spaces.

In Section 4 we use a simplification of the clustering technique to build a data structure (index) on a space of text documents. This data structure permits us to search for documents similar to a given one or to a text query, under the classical cosine similarity measure used in IR. The search technique exploits the clustering properties of the index structure by scanning the most promising clusters first. The approach is an approximation: with a very low probability of missing relevant documents, we can answer queries by making only a fraction of comparisons (distance computations) needed for an *deterministic* algorithm.

Finally, Section 5 resorts to a more general form of clustering: the data is partitioned according to an arbitrary property, not only to spatial closeness. This is used to partition a data set into subsets that behave different from each other in terms of their distances to the rest. The goal is again to build a faster index using information of the cluster structure of the data. Now the goal is to improve a deterministic algorithm by fine tuning the algorithm for each discovered cluster. In the example presented, different index parameters are used on each partition, each best suited to the local properties of each group. In a general perspective these indexes may have different local parameterizations of a single scheme, or even be completely different approaches, such as a an approximated approach in one partitions and an exact approach on a different partition. We demonstrate the effectiveness of the approach on an application to find the correct versions of misspelled words.

2 Our Clustering Method

Cluster and outlier detection are aimed at producing a model of the data. The primary objective of an application is to partition the data into a meaningful collection of sets. A partition, in the mathematical sense, is a collection of non-intersecting subsets whose union is the whole set. If the clustering defines a partition, then we will have a *crisp* clustering. If the subsets have non-empty intersection then we will have a *fuzzy* clustering. In this chapter we are interested in crisp clusters.

The data model obtained from cluster detection may be very sensitive to outliers. The outliers could be considered as "rare" data, not following the overall tendencies of the group. In order to design a robust clustering algorithm we need to properly define what a cluster is. This definition should follow some intuition and be computable in a reasonable amount of time. In addition, the clustering should be able to give the cluster of a given element without clustering all the data, and it should be able to detect outliers.

Two additional restrictions are of importance. The clustering should be carried out using only the distances between objects, and no new objects should be created. These restrictions are a fair way to mask the application level. All the domain knowledge will be encapsulated in the distance computation. The resulting clustering technique will be suitable for document retrieval, for browsing images, or for handling multimedia data in general.

Cluster and outlier detection is a classic problem of non-parametric statistics. Cluster analysis is the detection of groups that reduce the inter-group similarity and increase the intra-group similarity. In this section we explore several approaches based on graphs for cluster analysis. The strategy is to define the clustering property and then to identify a random graph capturing it. This generic algorithm will satisfy our restrictions and hence will be suitable for metric spaces.

The ultimate goal of any clustering algorithm is to postulate a model of the data. Different applications define different models and hence different objective functions, each of which will be optimized by a different clustering. This optimization goal function may be implicit or explicit. For exact indexing purposes (such as the application of Section 5), the implicit goal function is the number of distance computations performed to solve a proximity query. A good clustering will produce a partition that minimizes the number of distance computations performed at search time. For approximate searching (as described in Section 4), the goal is to optimize the tradeoff between accuracy of the result and number of distance evaluations. For classification purposes (described in Section 3) the goal will be just the accuracy of the classification.

Most clustering methods considered in the statistical literature and in statistical software, namely partitioning methods such as k-means and k-medians [JD88], and agglomerative methods such as dendograms [Har75], share a "two-stage" nature. First, by assigning different values to some parameter (for instance, k in k-means clustering), the algorithm produces a family of clustering structures, that is, a family of partitions of the original data into relatively homogeneous disjoint blocks. Second, the user must resort to some validation criterion, such as those studied in [MC85], or use some exploratory analysis of the results, in order to decide which is the "right" clustering structure. It is desirable to have clustering procedures that are "automatic", in the sense of providing the user at once with the amount and identity of the clusters present in the data. This automatic procedure will be appealing only if we may find a way to disaggregate those objects not belonging to any group. This feature will be crucial for data cleaning applications.

On the other hand, most classical clustering algorithms have quadratic or even higher complexity, because they must measure global properties involving every distance pair. Clustering is an inverse problem, just like interpolation. This means that it is not well defined in the mathematical sense unless we add additional constrains, called "regularity assumptions", which distinguish between good and bad solutions. A rather classic example of regularity conditions for a problem is the problem of fitting a curve to a set of discrete points, or to interpolate the data. An infinite number of solutions exist for this problem, but we select a single *smooth* curve as *the solution.* In detecting the cluster structure of multivariate data we must make some kind of regularity assumptions. Most of the proposed algorithms in the literature are based on heuristic regularity assumptions about the nature of the clusters and their shape. The user iterates several times, accepting some hypotheses and rejecting others, and each assumption leads to a different clustering of the data. A difficult problem remains: naming clusters, particularly in IR.

2.1 Clustering in Metric Spaces

We will focus on applications where the objects and their similarity define a *metric space.* This is formalized as a universe of valid objects $\mathbb{X}$ and a distance function d that satisfies the triangle inequality and is typically expensive to compute. We manage a finite data set $\mathbb{U} \subseteq \mathbb{X}$, of size n. The problem of clustering can then be defined as partitioning $\mathbb{U}$ into sets such that the intra-cluster distances are "small" and inter-cluster distances are "large". In several applications, clustering is used as a tool for answering proximity queries of the form (q, r), where $q \in \mathbb{X}$ and $r \in \mathbb{R}^+$. This query defines a *query ball* $(q, r) = \{x \in \mathbb{X}, d(q, x) \leq r\}$, and the goal is to retrieve the elements of the database that are inside the query ball, that is, $(q, r) \cap \mathbb{U}$. Other types of queries are possible, but we will focus on the simplest case. The aim of *metric space searching* is to preprocess the database $\mathbb{U}$ so as to build an index that answers queries (q, r) performing as few distance computations as possible [CNBYM01]. We will show that clustering $\mathbb{U}$ yields large performance improvements on the searching performance of the indexes.

In cluster analysis one often assumes that the data set comes from a d-dimensional real vector space $\mathbb{R}^d$. Some algorithms, like k-means, make heavy use of the coordinate information; assuming, for example, that new centroids can be created. If neither the coordinates of the data, nor explicit object creation is used, then the algorithm is likely to be extensible to metric data sets. Further care must be taken if one wishes to extend a clustering algorithm to massive data sets, as the algorithm must scale in performance as the input data size grows. It is important to note that the distance function could be very expensive to compute, and therefore the complexity must take into account this leading complexity term.

Recent work has been done on clustering metric spaces by directly using a mapping from the original metric space into a low dimensional vector space, and then clustering the data using a traditional technique or an ad-hoc variation. This approach was tested in [GRG$^+$99] using a generalization of the well known BIRCH clustering procedure [ZRL96], originally designed for vector spaces. The strategy is to map the metric space into a vector space using FastMap [FL95], and then to use BIRCH in the mapped space. To obtain a good clustering the mapping must be accurate. Unfortunately, FastMap is neither contractive nor expansive if the original space is not $\mathbb{R}^d$ with the Euclidean distance. Hence the distortion in the distances is unbounded. The algorithm proposed in [GRG$^+$99] is linear in time. Our clustering approaches will make a single pass over the data set to discover each grouping, hence their total complexity is at most quadratic.

2.2 Mutual k-Nearest Neighbor Graph Clustering Algorithm

The problem of identifying the clustering structure is hard. We may focus instead on the converse problem, identifying the *absence of clusters* in a sample data set. If we have a small amount of sample data, we cannot decide if they cluster together, since we have little implicit information about their relative proximity. So we begin by asking for a significant number of sample points. We believe that most readers will agree that if the data set is drawn from a uniform distribution, then the data set is cluster-free. This can be generalized to distributions bounded away from zero and from infinity, that are almost-uniform distributions. If the data set is cluster-free, we say there is a single cluster integrated by all the elements of the sample. If the data have a cluster structure, it is reasonable to expect the data to be almost-uniform inside each clusters. We use this observation to base our approach in the null hypothesis testing "the data is cluster free".

In [BCQY97], a technique is proposed to detect clusters and outliers based on a graph theoretical-approach. This technique has provable cluster-detection power under some regularity conditions: (a) the clusters to be detected correspond to different connected components in the support[2] of

[2]The support of a random variable x, with density function $f(x)$, is the set S such that

the sample distribution, and these connected components are at strict positive distance from each other, (b) the sample distribution has a density bounded away from zero and infinity on each component of the support and (c) the connected components of the support of the distribution are grid compatible[3]. We call a distribution with this properties a *regular* distribution.

We will call clusters the connected components of the support. Since it is assumed that connected components are at a strictly positive distance from each other, the algorithm can detect "crisp" clusters, as opposed to "fuzzy" clusters with overlapping support.

The technique just described postulates a graph, over random points in $\mathbb{R}^d$. The random graph has a controlling parameter τ for its density (e.g. the number of arcs for each node). Above certain threshold the graph is connected with high probability. The threshold for τ depends in turn on the underlying point distribution, the dimension of the space and the number of points. In short, we can always have a value of τ such that the graph will be connected with high probability if the sample points are drawn from a distribution that is bounded away from zero and from infinity and will not be connected if the points come from a distribution vanishing asymptotically in the support. If we have a sample data set coming from a regular distribution our goal will be to detect the clusters (or equivalently to detect the components of the support). Many graphs could be defined for clustering purposes under the above setup. In [BCQY97] they use the Mutual Nearest Neighbor graph, defined below. In this chapter we also describe the range-r graph.

The controlling parameter τ of the random graph depends on the characteristics of the probability space. The general technique relies on the proper estimation of the connectivity threshold of τ in the random graph. In general, the estimation could be two fold, a theoretical bound and a practical recipe based on a Montecarlo simulation. To find the threshold, the procedure is to generate particular instances of cluster-free datasets. In these datasets we observe the minimum value of τ such that the random graph is connected.

For example, we may choose uniformly distributed points in $\mathbb{R}^2$ and estimate the minimum value of the parameter τ to obtain a single connected component. After estimating τ we will be able to detect clusters (i.e.

$f(x) > 0$. We strength this condition by assuming $f(x) > \epsilon$ (bounded away from 0).

[3]See [BCQY97] for details, the essential idea is that the support accepts a discretization (it cannot be infinitely thin) and it is completely contained in a discrete grid, such that each grid cell has neighboring non-empty cells.

connected components) in arbitrary data, if the number of points and the dimension of the sample data, are both similar to the estimated values.

The technique described above can be applied to data sets with the same topological characteristics. For example the parameter τ for $\mathbb{R}^2$ and $n = 1000$ points in the sample may not be applied to data sampled from $\mathbb{R}^{23}$ and $n = 50,000$ points. Hence the controlling parameter τ will be a function of the space topology.

After this digression, the central question is: how to apply the procedure to general metric spaces? In this case we will need to find out an invariant in the sampled data and, for clustering, the parameter τ must be applied to data sets sampled from a probability space with the same invariant. We will use as invariant the intrinsic dimension of the data, defined as the minimum d such that the data can be embedded in $\mathbb{R}^d$ without distortion. This approach can be used as long as the random graph can be computed using only the distances among the sample data. Note, however, that in most cases computing the distances between sample elements is very expensive; hence the algorithm performance is measured in the number of distance computations. Moreover, if the data is *massive* (gigabytes to terabytes of data) it is advisable to use a technique that can work in secondary memory, because disk accesses could be far more expensive than distance computations.

Since the above embedding may be hard to find (as hard as finding the clustering structure), one may resort to hierarchical clustering, by moving the parameter τ to find the proper threshold. In this section we describe a clustering algorithm based on the *Mutual Nearest Neighbor Graph* and the *Range r Graph* respectively. This clustering procedure is able to discover clusters at different resolutions, and has subquadratic complexity measured in terms of the number of distance computations. To achieve this complexity the algorithm relies on the use of an indexing algorithm. This indexing can be implemented using any indexing data structure. A better indexing algorithm will lead to faster clustering, without impact on the quality of the clusters obtained.

2.2.1 The Clustering Algorithm

From an algorithmic point of view the procedure is to find the k nearest neighbors of a seed point s, the set $kNN(s) = \{s_1, s_2, \cdots, s_k\}$, and to include in the cluster of s those points $s' \in kNN(s)$ having the reciprocal property $s \in kNN(s')$; and then to proceed iteratively until no more points could be added to the current cluster, restarting with a new seed point until no more seed points are left to visit. This algorithm is shown in figure 2.2.1

with S being the set of points, and k the number of neighbors considered. The output is a partition of the sample data, and each partition element is either a cluster or an outlier.

```
MkNNCluster (k, s, S)

        let S'_s = {s' ∈ kNN(s) || s ∈ kNN(s')}
        return s∪ MkNNCluster (k, s'_1, S) ∪ ··· ∪ MkNNCluster (k, s'_k', S)

Partition (k, S)
        let S = {s_1, ··· , s_n}
        while(|S| > 0)
              output MkNNCluster(k, s' ← pick(S), S)
              S = S−MkNNCluster(k, s' ← pick(S), S)
        endwhile
```

Figure 1: The $MkNN$ clustering algorithm, k is the number of neighbors to consider.

Calling the procedure **Partition** of figure 2.2.1 will identify all the connected components of a graph defined implicitly, called the mutual k nearest-neighbor graph. A technical drawback in this approach, is to find out the correct value for k. It is clear that, as k increases, the number of connected components decreases; this gives a natural way to define a hierarchical structure. Using brute force for finding the k nearest neighbor of each site in the data set, takes $O(n^2)$ distance computations. In the next subsection we discuss in some detail how the above algorithm, and a family of algorithms based on similar assumptions, can be converted on a clustering detection procedure.

2.2.2 Connectivity Properties

The algorithm just described depends heavily on the parameter k, the number of neighbors to use. For $k = 0$ the partition consist in one point for each partition element. If $k = n - 1$ then we have surely one single set in the partition. For any number of neighbors k in between we will have a finer or a coarser partition. We are interested in finding the "right" number of clusters and we have as a parameter the number of neighbors, k, to use. If we cannot postulate any a priory knowledge neither about the type nor the shape of support of the distribution, we can always build a hierarchy of partitions.

With a large enough k we can compute the k-nearest neighbor graph of the data set and for each $k' \leq k$ we don't need any additional distance computation to find the k'-nearest neighbor. Each partition in the hierarchy is a coarsification of the preceding partition. In other words, if $k_1 \leq k_2$ then each equivalence class in the partition induced by k_1 is completely contained in one equivalence class of the partition induced by k_2. Once the hierarchy is postulated, the "right" number of clusters could be found using well known intracluster/intercluster stress measures.

If our goal is to produce an "automatic" clustering procedure, we need to find a priory the value of the parameter k; and bound the search for a particular class of distributions *inside* the connected components of the support. In particular we can restrict ourselves to vector spaces, or to metric spaces that can be embedded into a low dimensional vector space. It is not necessary to explicitly find the mapping, instead we can postulate a null hypothesis.

In [BCQY97] are proved both analytical and experimental bounds for k, depending on both the size of the sample and the dimension of the embedded vector space. distribution is regular enough. The bounds proved are not tight, but they postulate a compact support and a uniform distribution inside the clusters. Under this setup, the connectivity constant can be estimated using a Montecarlo approach. The idea is to estimate the minimum k such that there is only one cluster for a given sample size, obtained from a uniform distribution. The data obtained in the Montecarlo simulation is fitted to a model of the same family of the theoretical bound, of the form $k_d = a_d + b_d \log(n)$, having different results for each different intrinsic dimension d.

2.3 The Range r Graph

We propose an alternative procedure to detect clusters and outliers, based on the same foundations of the $MkNN$ graph. The central idea is to use *range queries* instead of k-nearest neighbor queries. Each site s will share an edge with s' in the G_r graph if $s' \in (s,r)_d$, with $(s,r)_d = \{s' || d(s,s') \leq r\}$.

With the above definition of G_r we can use algorithm 2.2.1 with essentially no modifications to partition the data into clusters and outliers. The graph G_r is simpler than $MkNN$, and not checking for symmetry speeds up the construction. In both approaches the use of an index to satisfy either k-nearest neighbor queries or range queries is advised to achieve subquadratic complexity.

RangeCluster (r, s, S)

  let $S'_s = \{s' \in (s, r)_d\}$
  return $s \cup$ RangeCluster $(r, s'_1, S) \cup \cdots \cup$ RangeCluster $(r, s'_{j'}, S)$

Partition (r, S)
  let $S = \{s_1, \cdots, s_n\}$
  while($|S| > 0$)
    output RangeCluster($r, s' \leftarrow pick(S), S$)
    $S = S-$RangeCluster($r, s' \leftarrow pick(S), S$)
  endwhile

Figure 2: The range clustering algorithm, r is the ball radius.

In algorithm **RangeCluster** presented in figure 2.3, we can see that the procedure is almost the same for our clustering strategy and the one obtained using the $MkNN$ graph. We proceed to obtain an adequate r according to our hypothesis that inside clusters the data distributes almost uniformly. The procedure is to make a Montecarlo simulation of the algorithm to obtain statistics on the minimum r such that there is only one cluster in the sample.

This algorithm performs on average $O\left(n^{1+\alpha}\right)$ distance computations (the operation of leading complexity), with $0 \leq \alpha \leq 1$ a constant depending on the intrinsic dimension of the sample data. The algorithm can detect outliers in the sample data and, if desired, it can produce a hierarchical structure (a dendogram) pointing to clusters at different resolutions.

2.3.1 Outliers, Equivalence Classes and Stability

The random graphs discussed in our scope are uniquely defined once the parameter τ (k and r, respectively) and the point set are fixed. The two examples we present in subsection 2.2.1 have additional advantages. One nice feature is that the random graph can be computed incrementally. In other words, we start with a single point and begin adding edges and vertices until no more vertices appear. We may think clusters as equivalence classes, and the incremental procedure ensures obtaining the same class for any representative. Hence, *the method is stable and deterministic.*

Certain points will be isolated in the MkNNG or the Range r graphs, forming a connected component of low cardinality. If the equivalence class of a point is the point itself (or a very small number of elements, compared with the complete data set), this equivalence class will be considered as an *outlier of the sample.* The discussion about the validity of this assumption is beyond the scope of this book. A common sense argument can be used however. A small equivalence class corresponds (according to our hypothesis) to a very small component of the support of the distribution. This implies either a very small probability of having points in that particular location, or having points where the support vanishes. In either case, the presence of these points does not follow the same general rule of the rest of the sample.

A final remark on the outlier issue. If *most* of the equivalence classes detected have a small cardinality, this naturally implies a small value (underestimated) of the parameter τ, and *not* the presence of "many" outliers.

2.4 Radius vs. Neighbors

The controlling parameter of both clustering approaches (Range and MkNN graphs), the number of neighbors k and the radius r, must be estimated beforehand. We face here a different problem, because we do not have a theoretical model on the searching radius, moreover we do not expect to obtain a fixed radius for every data set. The searching radius must scale with the density of the points.

The goal of our Montecarlo analysis is to find an invariant giving us a rule to estimate the radius of the RangeCluster procedure. We will have at hand only the sampled data, with no good information about the intrinsic dimension of the data. We can use only inter-point statistics to estimate the parameter r. If the particular case of spatial data (sampled from an Euclidean d-dimensional space) we can use intrinsic dimensionality estimators, but this does not apply to metric data from an arbitrary space.

To guide the search we ran the clustering algorithm in uniformly distributed data in the d-dimensional unitary cube. The goal is to estimate the minimum radius to obtain a single cluster. The variables of the experiment are the number of points and the dimension. We had to take into account that as the number of points increased the minimum inter point distance decreases, since the cube becomes more dense. Note that the number of points in a ball grows exponentially with the ball radius.

The relation found was the *average number of points* inside the $(s,r)_d$ ball. For a fixed dimension, independently of the number of points (the data density), we found that when the average cardinality of the $(s,r)_d$ ball is above certain threshold there is always a single cluster in our setup. The radius of the $(s,r)_d$ ball was variable, revealing small changes in the local density even in the uniformly distributed data.

The rule is then simply stated as searching with a radius such that the ball captures at least n_d points. If the data is very concentrated (with high local density) then the searching radius is small and conversely. This implies that we need inter-point statistics, to estimate the radius of the $(s,r)_d$ ball for the algorithm. Fortunately we can use statistics from the index construction to propose the appropriate searching radius, since the construction of the index uses inter point distances in the construction process. In other words it does not imply an overhead in the number of distance computations.

2.5 The Connectivity Parameters

We include the results of the Montecarlo simulation for the connectivity of the MkNN graph from [BCQY97] below, for low dimensional data. The R^2 column is the fitness value, high values are better. The sampled standard deviation $\hat{\sigma}_d$ is also shown.

dim	adjusted model	R^2	$\hat{\sigma}_d$
2	$4.341 + 0.430 \ln n$	0.979	1.313
3	$3.422 + 0.646 \ln n$	0.984	1.744
4	$2.651 + 0.924 \ln n$	0.983	2.266
5	$1.859 + 1.219 \ln n$	0.976	2.674
6	$0.268 + 1.687 \ln n$	0.994	3.092
7	$-0.770 + 2.019 \ln n$	0.989	3.364
8	$-2.453 + 2.474 \ln n$	0.994	3.666

Table 1: Adjusted models for k (as described) in the MkNN random graph (uniform data in $[0,1]^d$).

Once the intrinsic dimension of the data is estimated, we can use table 2.5 to compute the connected components of the MkNN graph, as described. The user will judge if the obtained equivalence classes correspond to clusters or outliers, based on the cardinality and the relative position of the data.

For the Range r graph a similar table can be obtained after the corresponding Montecarlo simulation. A threshold parameter τ must be found depending on the number of points, and the (intrinsic) dimension of the data. In this case the threshold parameter cannot be easyly stated in absolute numbers, since the inter-point distance is variable. If we fix the number

Dimension	2	3	4	5	6	7	8	9	10
Sites	4.37	5.07	5.51	6.05	6.22	8.03	7.78	8.59	10.32

Table 2: The minimum number of points to have a single cluster, in different dimensions.

of points in the sample, and regard the "minimum radius such that m points are captured" we will obtain a table similar to table 2. We can also fix the radius as a percentage of the minimum inter-point distance, the maximum inter-point distance, etc. The proper invariant should be selected according to the user needs.

2.6 Intrinsic Dimension

The intrinsic dimension of the data is an important invariant in metric space samples. Since one cannot assign directly coordinates or a dimension notion to the data, we have to resort to vector spaces where the notion of dimension is properly defined. If the intrinsic dimension of the data is known the clustering structure could be detected. Conversely, if the cluster structure is known, the intrinsic dimension could be found.

Both arguments are valid. When discovering the clustering structure of a sample we are postulating a particular distribution of the data. This implies that we have an inverse problem, since the intrinsic dimension, and distribution are unknown, we have to fix one of them. The clustering procedure described could be used also as an intrinsic dimensionality estimator. The Range r graph, or unit distance graph, and the MkNN graph could be used as an intrinsic dimensionality estimator. The procedure will be the converse of the clustering procedure. We discover the clustering structure using a hierarchical approach. After this we obtain (from the appropriate table) the intrinsic dimension corresponding to the value of the parameter, and the number of points used.

3 Morphological Stemming and the Holomorphic Distance

3.1 Motivation

The problem of conflating words sharing a common root is an important task in information retrieval. The standard solution for stemming is a rule based approach inherited from Porter's algorithm. This type of solution closely follows the possible variations of a single stem into all its morphological instances. This technique suffer from several drawbacks, being the main problem the presence of *noise* in the pattern, i.e. spelling errors. When a word is misspelled then finding its correct stem by following rules is at least risky. A single edit error in a word may trigger the wrong rule and that will cause a sure mistake in the conflation.

In this section we describe a different alternative for stemming and conflation. This technique has been proposed in [Cha02]. The idea is to use the so called holomorphic distance to detect words with similar stems, and then to use a clustering algorithm to isolate clusters. The input is a dictionary of words and the output is a partition of the dictionary into words sharing a stem.

Traditional sequence comparison is done by carefully aligning the formant symbols of a sequence to obtain the minimum number of atomic operations (insert, delete, substitute) to convert one sequence into another. The de-facto choice for comparing sequences has been the Levenshtein or *edit distance* and variations obtained by adding or inhibiting atomic operations.

This naturally implies that the distance takes into account only local information. Moreover, our sequence comparison does not use the context of strings, as the expert knowledge in some very focused fields cannot be meaningfully mapped into insertion/deletions/substitutions or other operations, mainly because the similarity perceived by the expert is not local.

The holomorphic distance for sequences is not based on the edit distance for words. The idea is to extract a number of features of the sequence (in this case a word in a document), and to use the features to build a vector, in the same fashion that is done for documents in IR. Below we give a more detailed description of the distance.

For information retrieval it is often desirable to conflate all the variations of a word into a single stem. This operation conflates, for example, `house`, `housing`, `houses` into the single stem `hous`. In this way, when the user wants to query the system with any of the variations of the word, all the documents containing the same stem will be retrieved. In a regular language (like English) there will be no problem in building a stemmer using the Porter's algorithm. In a Semitic or romance language the stem may not be a prefix of the word, and the construction of a stemmer using the Porter's approach may not lead to an efficient solution. Other problem often faced in either language is querying or indexing misspelled words. In this case even a perfect stemmer would fail to conflate to the proper stem. A *morphological* stemmer, on the other hand, will make use of the word structure to guide the conflation process. To this end it would be necessary to assign a small distance to all the words with the same structure and a large distance to words with different structure.

3.2 The Holomorphic Transformation

The holomorphic transformation generalizes the notion of similarity developed for documents. The general idea is to decompose the original sequence in a number of sub-sequences using a number of fixed rules. This rules will be *feature* extractors of the sequences. The rule will be to obtain sub-sequences when a feature extractor is applied to a sequence. If Σ is the alphabet, and $s \in \Sigma^*$ is a sequence, then a function $\Gamma : 2^{\Sigma^*} \to 2^{\Sigma^*}$ is a *feature extractor*. We put no constraints to the type of constructions we allow for extracting features. Any function will fit our needs.

The limits imposed will be (if any) the computational complexity to obtain the transformation, and the a priori knowledge of the designer.

The *holomorphic transformation* will be a function $\mathcal{H} : \Sigma^\dagger \to I\!R^m$ with m the size of the context $[\Gamma^\dagger, \mathbb{S}]$. This transformation depends heavily on the set of feature extractors. The context will be the vehicle to introduce domain knowledge to the problem of sequence comparison. Note that once a sequence is transformed using the holomorphic transformation, we can use the cosine distance to compare the corresponding vectors.

The context is the generalization of the vocabulary of a document collection. Under this point of view, the distance between a given pair of sequences will depend on the context. The distance will not be absolute any longer. Note also that the positional information may be lost after the holomorphic transformation. This is the case of transforming documents into vectors. If we want to recover the positional information we need to be very careful with the design of the feature extraction functions. In the following subsection we will discuss some examples of this.

3.3 A Morphological Stemmer Using Clustering

We can use the holomorphic distance to build a morphological stemmer. For a romance language the stem of a word is almost surely certain prefix, and the syntactical category is certain suffix. If we use q-grams, prefixes and swaps (Γ_q^g, Γ_j^p and $\Gamma_j^{2!}$), and carefully select the relative weights of the individual features, we are in position of measure the morphological closeness for word pairs. Tables 3 and 4 are examples of the closeness using this approach. The idea of the distance between the words is to find the

	H. non weight	H. weight	Edit distance
1	apetitosa	apetitosa	apetitosa
2	apetito	apetitoso	apetitoso
3	apetitoso	apetito	aceitosa
4	apetite	apetite	apestosa
5	apetitiva	apetitiva	apetitiva
6	apetitivo	apetitivo	apetito
7	apetible	apetible	aceitoso
8	apetecer	apetecer	acetoso
9	apetecedor	apea	alentosa
10	apetencia	apetencia	aparatosa

Table 3: The 10 nearest neighbors of the Spanish word *apetitosa* with two different weighted holomorphic distances, and the edit distance. The edit distance reports words with different meaning (3-4, 7-10), while the holomorphic distance provides very strong closeness (except 9 in the weighted case).

maximum prefix, allowing misspellings, between two words. This maximum prefix is weighted by the presence of similar prefixes in the corpus.

The complete algorithm for stemming using the holomorphic transformation has two stages. First, we find the clusters of the corpus using the holomorphic distance. These clusters coincide with the words conflated for sharing the same stem. This defines equivalence classes by morphological closeness. In the second stage we query the corpus and find the closest equivalence class, tp report the same representative for each equivalence class member. This representative may be some normal form of the stem (e.g. the verb in infinitive) or the largest common stem in the equivalence class.

A related problem in information retrieval and computational linguistics, is syntactical tagging. Here the problem consist in assigning a label to the word, indicating if it is a verb, adverb, noun, etc. A similar scheme as the one described for stemming may be followed to tag words.

4 Clustering for Approximate Proximity Search

In this section we describe an approximate proximity search algorithm based on a clustering data structure. The clustering method we use is in the same

	H. non weight	H. weight	Edit distance
1	dettagli	dettagli	dettagli
2	dettaglio	dettaglio	dettaglio
3	dettagliato	dettagliata	dettai
4	dettagliati	dettagliate	dentali
5	dettagliate	dettagliati	deragli
6	dettagliare	dettagliato	dettami
7	dettagliata	dettagliare	dettati
8	dettagliante	dettagliante	dettavi
9	dettaglianti	dettaglianti	rettali
10	dettagliatamente	detta	battagli

Table 4: The 10 nearest neighbors of the Italian word *dettagli* with two different weighted holomorphic distances, and the edit distance, with results similar to the Spanish case.

spirit of that described in Section 2.2, although simplified.

The *List of Clusters* data structure [CN00] is a list of "zones". Each zone has a center c and a *covering radius* $cr(c)$, defined as the maximum distance between c and any element of the zone. A parameter of the structure is the number m of elements inside each zone (counting the center).

To build the structure, a first center $c \in \mathbb{U}$ is chosen at random, and its $m-1$ nearest neighbors are found. A zone is made up of c and these $m-1$ neighbors. Its covering radius $cr(c)$ is defined as the distance to the $(m-1)$-th neighbor. Once this first zone, I, is determined, we remove the elements of the set and build the rest of the list over $E = \mathbb{U} - I$. The next center is chosen as the element that maximizes the sum of distances to the previous centers.

The construction process returns a list of triples (c_i, cr_i, I_i) (center, radius, zone elements). Figure 3 (left) shows the pseudocode for the construction algorithm. On the right we show an example with three zones. Note that this data structure is asymmetric, because the first center chosen has preference over the next centers in case of overlapping balls. The brute force algorithm for constructing the list takes $O(n^2/m)$ distance computations, but it can be improved using auxiliary data structures to find the neighbors.

To solve a query (q, r), $d(q, c)$ is computed, reporting c if it is within the query ball. Then, we search exhaustively inside I only if $d(q, c) - cr(c) \leq r$, because otherwise the zone with center c cannot have intersection with the

BuildList (Set of objects $\mathbb{U}$)

if $\mathbb{U} = \emptyset$ **then return** empty list
Select $c \in \mathbb{U}$
$I \leftarrow$ the m elements in $\mathbb{U}$ closest to c
$cr \leftarrow \max_{u \in I} d(c,u)$
$E \leftarrow \mathbb{U} - I$
return $(c, cr(c), I)$: BuildList(E)

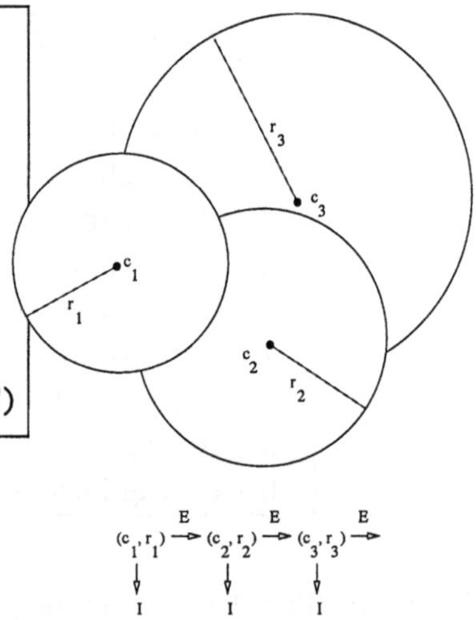

Figure 3: On the left, construction algorithm of List of Clusters. Note that I includes c. The operator ":" is the list concatenation. On the right, a graphical example.

query ball. Also, given the asymmetry of the data structure, we can prune the search in the other direction: if the query ball is totally contained in $(c, cr(c))$, that is, if $cr(c) - d(q,c) < r$, then we do not need to consider E, because all the elements that are inside the query ball have been inserted in I. Figure4 depicts the search algorithm.

SearchList (List L, Query q, Radius r)

if $L = \emptyset$ **then return** /* empty list */
Let $L \leftarrow (c, cr(c), I) : E$
Evaluate $d(c,q)$
if $d(c,q) \leq r$ **then** report c
if $d(c,q) \leq cr(c) + r$ **then** search I exhaustively
if $d(c,q) > cr(c) - r$ **then** SearchList(E, q, r)

Figure 4: Search algorithm on List of Clusters.

The search cost for this data structure has a form close to $O(n^{\alpha})$ for some $0.5 < \alpha < 1.0$ [CN00].

4.1 The Vector Model for Information Retrieval

In IR [BYRN99] a *document* is defined as the *recovery unit.* This can be a paragraph, an article, a Web page, etc. The classical models on IR consider that every document is described by a representative set of keywords called *terms.* A term is a word whose semantics help describe the principal topics of a document.

The most popular of these models, the *vectorial model*, considers a document as a t-dimensional vector, where t is the total number of terms in the system. Each vector coordinate i is associated to a term of the document, whose value corresponds to a positive *weight* w_{ij} if that term belongs to the document or zero if not. If $\mathbb{D}$ is the set of documents and d_j is the j-th document in $\mathbb{D}$, then $d_j = (w_{1j}, w_{2j}, \ldots, w_{tj})$.

In the vectorial model, a query object q can be a list of a few terms or even a whole document. The *similarity* between a document d and a query q is taken as the similarity between vectors $\vec{d_j}$ and $\vec{q}$, and quantified as the cosine of the angle between the two vectors:

$$sim(d_j, q) = \frac{\vec{d_j} \cdot \vec{q}}{|\vec{d_j}| \times |\vec{q}|} = \frac{\sum_{i=1}^{t} w_{ij} \times w_{iq}}{\sqrt{\sum_{i=1}^{t} w_{ij}^2 \times \sum_{i=1}^{t} w_{iq}^2}} \quad (1)$$

where w_{iq} is the weight of the i-th term of the query q. The weights of the terms can be calculated using *tf-idf* schemes:

$$w_{ij} = f_{i,j} \times \log\left(\frac{N}{n_i}\right) \qquad\qquad f_{i,j} = \frac{freq_{i,j}}{\max_{\ell=1\ldots t}(freq_{\ell,j})}$$

where N is the total number of documents, n_i is the number of documents where the i-th term appears, and f_{ij} is the *normalized frequency* of the i-th term: $freq_{i,j}$ is the frequency of the i-th term in d_j.

If we consider that the documents are points in a metric space, then the problem of searching for documents similar to a given query is reduced to a proximity search in the metric space. Since $sim(d_j, q)$ is only a similarity function, we will use the angle between vectors $\vec{d_j}$ and $\vec{q}$, $d(d_j, q) = \arccos(sim(d_j, q))$, as our distance metric, so $(\mathbb{D}, d)$ will be our metric space.

Despite of this clear link, metric space techniques have seldom been used for this purpose. One reason is that the metric space of documents has a very high dimension, which makes any exact search approach unaffordable. In fact, there is no exact algorithm that can avoid an almost exhaustive search on the database to answer proximity queries in this particular metric space.

The standard practice is to use an inverted index to find the documents that share terms with the query and then search this set of candidate documents by brute force. This technique behaves well on queries of a few terms, but it becomes impractical when queries are documents. This is required, for example, in relevance feedback scenarios.

In the latter case, it is customary to use heuristics, because the definition of relevance is already fuzzy. In most cases, finding some good answers is as good as finding them all, particularly when the search cost is drastically reduced. This is a case where metric space search with approximation algorithms would be of great value, as it is much better to use an approximation where the quality and degree of exactness is well understood, than a heuristic which resists analysis.

4.2 Techniques for Approximate Proximity Searching

We describe a search technique where the user can tune between efficiency and quality of the answer. The maximum number of distance computations we can perform is fixed and denoted by *quota*. Once *quota* has been reached, no more elements can be considered. The search algorithm is approximated in the sense that some relevant elements can be missed because we could not find them before we exhausted the *quota*. Hence, it is crucial to use the allotted *quota* efficiently, that is, to find as soon as possible as many elements of the result as possible.

The technique described in this section is called *ranking of zones* [BN02]. The idea is to sort the zones of the List of Clusters in order to favor the most promising, and then to traverse the list in that order. The sorting criterion must aim at quickly finding elements that are close to the query object. As the space is partitioned into zones, we must sort these zones using the information given by the index data structure.

We compute $d(q,c)$ for every center, and estimate how promising a zone is using only $d(q,c)$ and $cr(c)$ (which is precomputed). One not only would like to search first the zones closer to the query, but also to search first the zones that are more compact, that is, the zones which have smaller covering radii (since all the zones have the same number of elements). Some zone ranking criteria are (all in increasing order, see Figure5):

- $d(q,c)$: the distance from q to each zone center.
- $cr(c)$: the covering radius of each zone.

- $d(q,c)+cr(c)$: an upper bound of the distance from q to the farthest element in the zone.
- $d(q,c)-cr(c)$: a lower bound of the distance from q to the closest element in the zone.
- $\beta(d(q,c)-cr(c))$: what we call *dynamic beta*.

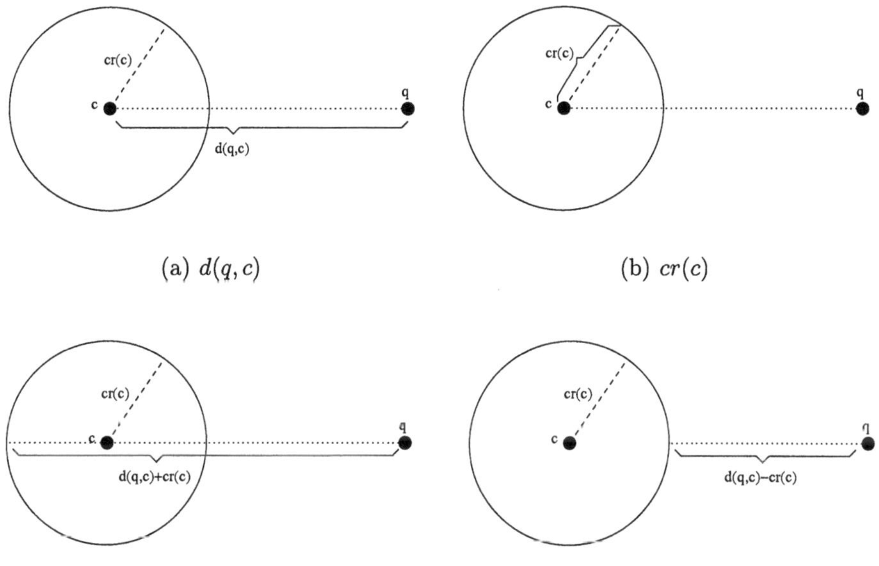

(a) $d(q,c)$ (b) $cr(c)$

(c) $d(q,c)+cr(c)$ (d) $d(q,c)-cr(c)$

Figure 5: Some zone sorting criteria.

The first two techniques are the simplest ranking criteria. The third technique aims to search first those zones that are closer to q and also are compact. The fourth technique uses the lower bound of the distance between the query object and any element of the zone, given by the index structure.

If factor β is fixed, then the last technique is equivalent to the criterion $d(q,c)-cr(c)$, because the ordering is the same. However, instead of using a constant factor $\beta \in [0..1]$, we use a *dynamic factor* of the form $\beta = 1/(1.0 - \frac{cr(c)}{mcr})$, where mcr is the maximum size of the covering radius of all zones. This implies that we reduce more the search radii in zones of larger covering radii.

4.3 Experimental Results

Figure 6 shows the results of experiments on a subset of 25,000 documents from *The Wall Street Journal* 1987–1989, from TREC-3 [Har95]. We compare the approximate algorithm using different ranking criteria. We used clusters of $m = 10$ elements and show queries with search radii that return, on average, 9 and 16 documents from the set. For example, on the left we see that, using the criterion $d(q, c) + cr(c)$, we can retrieve 90% of the results using 10,000 distance computations, that is, examining 40% of the space. We recall that all the exact algorithms require examining almost 100% of this space.

The results show that the approximate algorithms can handle well this space, and that the best criteria were $cr(c)$ and *dynamic beta*. We could retrieve more than 99% of the relevant objects while traversing merely a 17% of the database. This is the first feasible metric space approach to this long standing problem.

To show up to which point the concept of clustering has been essential for this good result, we consider a tempting declustering idea. Instead of sorting the clusters according to how promising the center looks, let us sort all the elements separately according to the extra information given by their distance to the center. For example, the lower bound criterion groups all the elements of a zone of center c under the value $d(q, c) - cr(c)$. If we knew that an element u in this zone is at distance $d(c, u)$ of its center, we could refine its estimation to $d(q, c) - d(c, u)$. This, however, performs worse than the original method. We conjecture that the reason is that we lose valuable clustering information when we rank each element separately.

Finally, we note that this data structure reminds the clustering techniques based on nearest neighbors covered in previous subsections, but it is more rough and simple. It is likely that better results can be obtained with a more sophisticated clustering algorithm.

5 Clustering for Metric Index Boosting

In this section we explore a different alternative for clustering. Instead of directly detect the data grouping by distance, we will detect the clusters in a different domain. For indexing purposes we are interested in characterizing the intrinsic difficulty of a given dataset. A step towards this is by detecting segments of the dataset which are more difficult to index than others. Here we propose a particular procedure obtaining clusters of data which are *not necessarily* close to each other in the distance domain. This can be put in

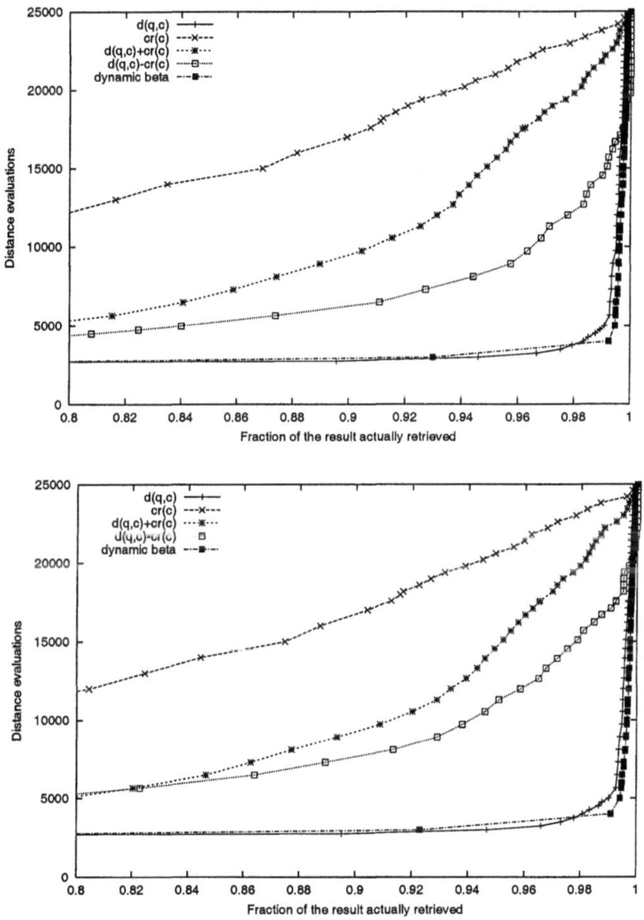

Figure 6: Comparison among different criteria in a document space, retrieving on average 9 elements (top) and 16 elements (bottom).

the following terms, for clustering we will be interested in minimizing the intra cluster distance $d(x, y)$, while maximizing the inter cluster distance $d(x, y)$. For our version of clustering we are interested in a different property, namely to minimize the intra cluster *measure* $f(x, y)$ and hence to maximize the inter cluster measure. The goal is to split the data in difficult-to-index and easy-to-index sets. We begin proposing a two way split, but this can be generalized to m-way splits as well.

Most indexing algorithms for proximity searching have tuning parameters. These parameters allow one to balance construction time, memory usage and search time, adapting the performance of the index depending on the characteristics of the data. The most relevant feature in a metric data set is how the data is distributed. Finding the underlying structure of a data set is very useful to design an indexing algorithm. In particular, knowing how the elements are clustered in the metric space help us identify the hardest region to be searched. Once the regions are categorized as easy, medium or hard for searching, we can locally tune the parameters for each region. Moreover, we can build independent indexes for each region, and search each index separately at search time. This has proven to be more efficient than using global parameters and a single index. Another application of the local parameterization technique could be, for example, to index one part of the database with an exact searching index and another part with an approximated index that will give good answers almost all the time.

In this section we use the data distribution to segment the database into just two parts: the hardest to be searched and the rest. This can be generalized to a finer partition, but we content ourselves with illustrating the technique. Notice that this does not correspond any more to a traditional clustering where we aim at grouping data that is spatially close. Rather, we group data that share some common properties. This can be seen as grouping data that are close after applying a change of domain.

One way of visualizing the data distribution is by using distance histograms. Given a metric space $(\mathbb{X}, d)$ and an element $p \in \mathbb{X}$, the *local histogram* with respect to the reference point p is the distribution of the distances from p to all the elements $x \in \mathbb{X}$. The local histogram can be very different from the *global histogram* of all the distance pairs in $\mathbb{X}$. However, if several local histograms are similar, then we can predict the behavior of the global histogram of a data set $\mathbb{U} \subseteq \mathbb{X}$.

One of the main difficulties in metric space searching is the so-called *curse of dimensionality*. Some metric spaces (called "high dimensional") have a very concentrated histogram, with small variance and typically large mean. This means that random pairs of distances are very similar from each other, or alternatively, that from the point of view of a given element p, all the others are more or less at the same distance. All indexing methods are based on precomputing some distances and inferring lower bounds on other distances [CNBYM01]. For example, if the index has precomputed $d(p,u)$ and, when searching for (q,r), we compute $d(p,q)$, then we know by the triangle inequality that $d(q,u) \geq |d(p,u) - d(p,q)|$, so we can discard u without ever computing $d(q,u)$ if it turns out that $|d(p,u) - d(p,q)| > r$. However, this (and any other attempt to avoid computing $d(q,u)$ for every u) becomes useless if the space is high dimensional, since in this case $d(p,u)$ will be very close to $d(p,q)$. However, for those elements u far away from the central region in the local histogram of p, the reference point p can be a good tool to discard them.

If a group of elements is at the same time in the central region of the histograms of several reference points, then those elements represent a subset where searching is inherently difficult. We call this group of elements the *hard kernel* of the space and denote it $hk(\mathbb{X}, d)$. The remaining elements belong then to a *soft kernel* denoted by $sk(\mathbb{X}, d)$. The idea is then to index and search separately the hard and soft kernels. That is:

- Partition the data set $\mathbb{U}$ into $hk(\mathbb{U}, d)$ and $sk(\mathbb{U}, d)$.
- Index separately $hk(\mathbb{U}, d)$ and $sk(\mathbb{U}, d)$.
- Solve (q,r) in $\mathbb{U}$ by searching $hk(\mathbb{U}, d)$ and $sk(\mathbb{U}, d)$ separately.

Detecting $hk(\mathbb{U}, d)$ is simple. We just intersect the central regions of local histograms for several different reference points p. After finding $hk(\mathbb{U}, d)$, we have that $sk(\mathbb{U}, d)$ is the complement.

Figure 7 describes the detection process of the hard kernel of a data set $\mathbb{U}$. The parameter s is the fraction of elements that should belong to the hard kernel. The parameter *cr* is the cutting radius used to delimit the central region in the local histogram of the reference point p. The idea is to take the elements surrounding the median of the histogram.

To test this algorithm we used a data structure for metric spaces called GNAT, explained next.

5.1 GNATs

GNATs (Geometric Near-neighbor Access Trees [Bri95]) are m-ary trees built as follows. We select, for the root node, m centers $c_1 \ldots c_m$, and define $\mathbb{U}_i = \{u \in \mathbb{U}, d(c_i, u) < d(c_j, u), \forall j \neq i\}$. That is, $\mathbb{U}_i$ are the elements closer

```
Compute_hk(Set of objects 𝕌, Fraction s, Radius cr)

hk(𝕌, d) ← 𝕌
Choose a point p ∈ 𝕌
while |hk(𝕌, d)| > s · |𝕌| do
    m ← median{d(p, u), u ∈ 𝕌}
    hk(𝕌, d) ← hk(𝕌, d) ∩ {x ∈ 𝕌, d(x, p) ∈ [m − cr, m + cr]}
    if 𝕌 ∩ hk(𝕌, d) ≠ ϕ
     Choose a point p ∈ 𝕌 − hk(𝕌, d)
    else
     Choose a point p ∈ 𝕌
return hk(𝕌, d)
```

Figure 7: Algorithm that finds $hk(\mathbb{U}, d)$

to c_i than to any other c_j. From the root, m children numbered $i = 1 \ldots m$ are built recursively as GNATs for $\mathbb{U}_i$. Figure 8 shows a simple example of the first level of a GNAT.

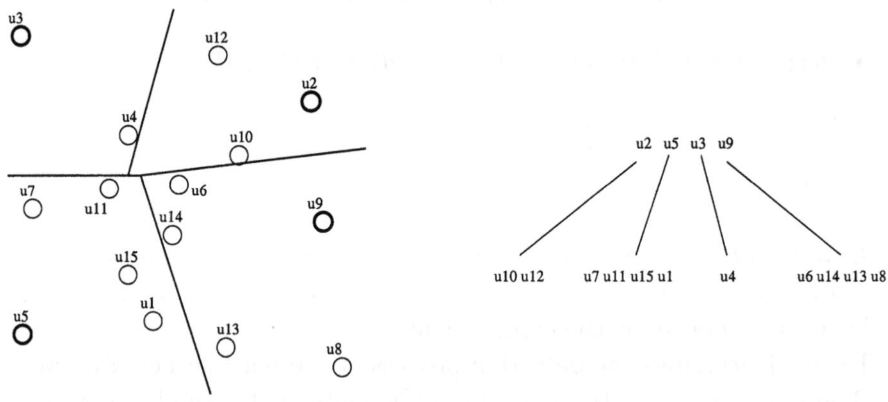

Figure 8: Example of the first level of a GNAT with $m = 4$.

The GNAT stores at each node an $O(m^2)$ size table $range_{ij} = [\min_{u \in \mathbb{U}_j}(c_i, u), \max_{u \in \mathbb{U}_j}(c_i, u)]$, with minimum and maximum distances from each center to each class. The tree requires $O(nm^2)$ space and is built in close to $O(nm \log_m n)$ time.

At search time, the query q is compared against some center c_i and we discard any other center c_j such that $d(q, c_i) \pm r$ does not intersect $range_{i,j}$, since all the subtree $\mathbb{U}_j$ can be discarded by the triangle inequality. The process is repeated with random centers until no one can be discarded.

The search then enters recursively into each non discarded subtree. In the process, any center close enough to q is reported.

The performance of the GNAT heavily depends on the arity of the tree. The best arity is different for each metric space, and even for different subsets of the data set. In particular, the hard kernel happens to require a large arity, while the soft kernel is searched better with a smaller arity tree. Hence, we will illustrate our technique by choosing different arities for hk and sk.

5.2 Experimental Analysis

We experimented with a metric space of strings under the *edit distance* (also called "Levenshtein distance"). This function is discrete and computes the minimum number of characters that we have to append, change and/or delete from one word to obtain the other. This distance has applications in information retrieval, signal processing and computational biology.

We have used a Spanish dictionary of 86,061 words, and experimented with different values for s and cr. The experimental setup consist in finding the best arity for the whole dictionary, and then splitting it in several ways to find a good proportion of soft/hard kernel tuning the individual arities of the proportions. Each combination is compared against the best arity for the whole dictionary. With this we ensure a fair comparison, since we compete with the best possible tuning of the GNAT for the whole dictionary, against individual tuning for the hard/soft kernels.

We chose 500 random words from the dictionary and searched for them using distance radii $r \in 1 \ldots 4$. For each search we computed the ratio of the search cost using separate indexes versus one global index. A value of 1 indicates the same performance in either approach, while a smaller value implies that the cluster-based approach is better than the standard one. Each point in the graph corresponds to the average measure of 500 queries, hence we expect a low variance for this measure.

Figure 9 shows the tuning of the GNATs, and we observe that an arity of 256 is the best we can do for low selectivity queries, while an arity of 128 is the best for high selectivity queries. We *must* choose a given arity if we will use only one index for all type of queries. We chose an arity of 256 to compare with the clustered approach.

For each cutting radius, and each proportion we tested all the combinations of arities for the cutting proportion. In figure 10 we show an example of such an experiment, for a cutting radius of 2, and a proportion of 0.4 in the hard kernel and 0.6 in the soft kernel and a fixed arity of 128 in the hard kernel, to find the best arity for the soft kernel. This experiment was

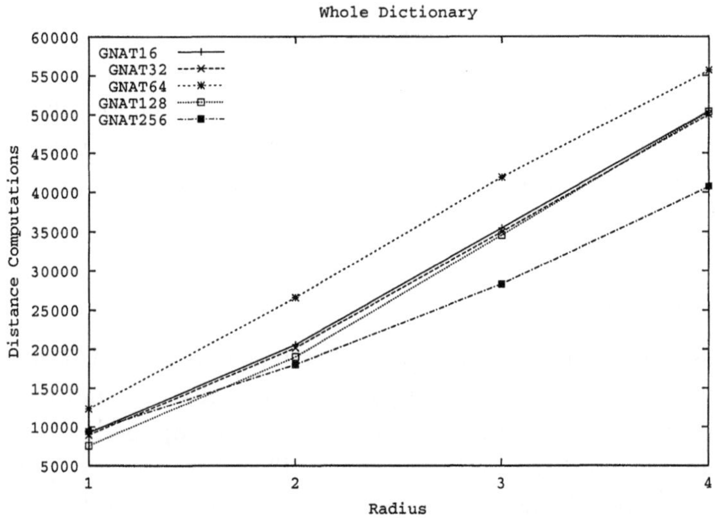

Figure 9: Tuning the GNAT. We try to find the best arity for the GNAT and found that this number is not the same for low/high selectivity queries. The best is GNAT-256 for most of the radii.

exhaustive in all the possible combinations.

In figure 11 we show how the different proportions are compared to the original dictionary. We observe a systematic improvement for high selectivity queries, while the improvement proportion is smaller in low selectivity queries. This is a natural consequence of the non-monotonicity of the tuning of the GNAT, but it is not explained solely by this, since for metric range queries of radius 2, both arities 128 and 256 in figure 9 have the same performance. The clustered index is better even in this case. We also observe that a balanced partition is the best choice, using an arity of 128 in each segment.

The example we have presented can be improved in a number of ways, for example by partitioning the data into more than two clusters, or by building a cluster hierarchy. More improvements can be expected by using different clustering strategies. The use of local histograms is a fast technique, but more costly techniques may produce a better segmentation.

Acknowledgements

We acknowledge the support of CYTED Project VII.19 RIBIDI. The first and last author also acknowledges the support of the Center for Web Re-

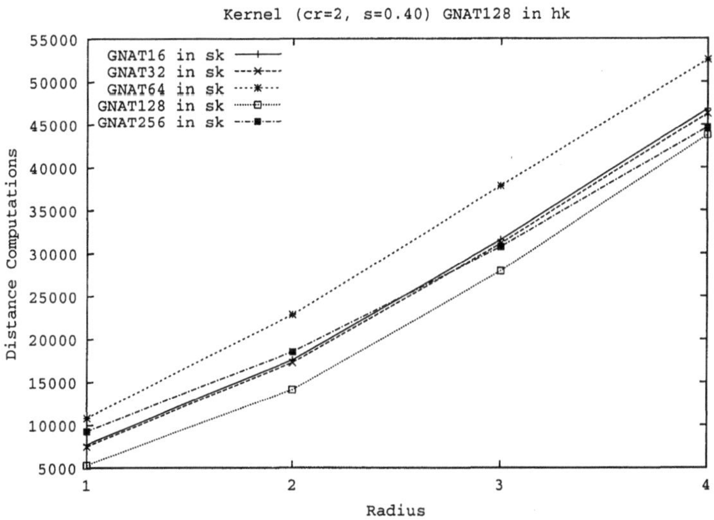

Figure 10: For each cutting radius and kernel proportion we selected the best combination of arities.

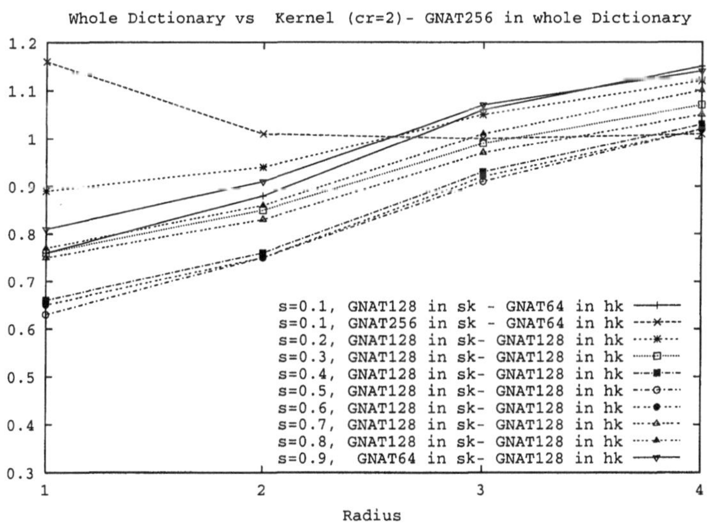

Figure 11: Once selected the best combination for each kernel proportion we compare it with the best-tuned GNAT. A balanced partition with arity 128 seems to be the best choice.

search, Milenium Research Initiative, Mideplan, Chile.

References

[BCQY97] M. R. Brito, E. L. Chávez, A. J. Quiroz, and J. E. Yukich. Connectivity of the mutual k-nearest-neighbor graph in clustering and outlier detection. *Statistics & Probability Letters*, 35:33–42, 1997.

[BN02] B. Bustos and G. Navarro. Probabilistic proximity searching based on compact partitions. In *Proc. 9th International Symposium on String Processing and Information Retrieval (SPIRE'02)*. LNCS, Springer-Verlag, 2002.

[Bri95] S. Brin. Near neighbor search in large metric spaces. In *Proc. 21st Conference on Very Large Databases (VLDB'95)*, pages 574–584, 1995.

[BYRN99] R. Baeza-Yates and B. Ribeiro-Neto. *Modern Information Retrieval.* Addison-Wesley, 1999.

[Cha02] E. Chávez. Knowledge based distances for sequence comparison. In *Sistemi Evoluti per Basi di Dati (SEBD)*, pages 34–47, 2002.

[CN00] E. Chávez and G. Navarro. An effective clustering algorithm to index high dimensional metric spaces. In *Proc. 7th International Symposium on String Processing and Information Retrieval (SPIRE'00)*, pages 75–86. IEEE CS Press, 2000.

[CNBYM01] E. Chávez, G. Navarro, R. Baeza-Yates, and J. Marroquín. Proximity searching in metric spaces. *ACM Computing Surveys*, 33(3):273–321, 2001.

[FL95] C. Faloutsos and K. Lin. Fastmap: a fast algorithm for indexing, data mining and visualization of traditional and multimedia datasets. *ACM SIGMOD Record*, 24(2):163–174, 1995.

[GRG+99] V. Ganti, R. Ramakrishnan, J. Gehrke, A. L. Powell, and J. C. French. Clustering large datasets in arbitrary metric spaces. In *International Conference of Data Engineering*, pages 502–511, 1999.

[Har75] J Hartigan. *Clustering Algorithms.* John Wiley & Sons, New York, 1975.

[Har95] D. Harman. Overview of the Third Text REtrieval Conference. In *Proc. Third Text REtrieval Conference (TREC-3)*, pages 1–19, 1995. NIST Special Publication 500-207.

[JD88] A.K. Jain and R.C. Dubes. *Algorithms for Clustering Data.* Prentice Hall Advanced Reference Series, 1988.

[MC85] G. W. Milligan and M. C. Cooper. An examination of procedures for determining the number of clusters in a data set. *Psychometrica*, 50:159–179, 1985.

[Zha71] C. T. Zhan. Graph theoretical methods for detecting and describing gestalt clusters. *IEEE Transactions on Computing*, 20:68–86, 1971.

[ZRL96] T. Zhang, R. Ramakrishman, and M. Livny. Birch: An efficient data clustering method for very large databases. In *ACM SIGMOD International Conference on Management of Data*, pages 103–114, 1996.

Clustering and Information Retrieval (pp. 35-82)
W. Wu, H. Xiong and S. Shekhar(Eds.)

Techniques for Clustering Massive Data Sets

Sudipto Guha
Department of Computer Information Sciences
University of Pennsylvania, Philadelphia, PA 19104
E-mail: `sudipto@cis.upenn.edu`

Rajeev Rastogi
Bell Laboraties,
Lucent Technologies, Murray Hill, NJ 07974
E-mail: `rastogi@research.bell-labs.com`

Kyuseok Shim
School of Electrical Engineering and Computer Science
Seoul National University, Kwanak P.O. Box 34
SEOUL 151-742, KOREA
E-mail: `shim@ee.snu.ac.kr`

Contents

1 Introduction

The wealth of information embedded in huge databases belonging to corporations (e.g., retail, financial, telecom) has spurred a tremendous interest in the areas of *knowledge discovery* and *data mining*. Clustering, in data mining, is a useful technique for discovering interesting data distributions and patterns in the underlying data. The problem of clustering can be defined as follows: given n data points in a d-dimensional metric space, partition the data points into k clusters such that the data points within a cluster are more similar to each other than data points in different clusters.

1.1 Shortcomings of Traditional Clustering Algorithms: Numeric Attributes

Existing clustering algorithms can be broadly classified into *partitional* and *hierarchical* [46]. Partitional clustering algorithms attempt to determine k partitions that optimize a certain criterion function. The square-error criterion, defined below, is the most commonly used (m_i is the mean of cluster C_i).

$$E = \sum_{i=1}^{k} \sum_{p \in C_i} \|p - m_i\|^2.$$

The square-error is a good measure of the within-cluster variation across all the partitions. The objective is to find k partitions that minimize the square-error. Thus, square-error clustering tries to make the k clusters as compact and separated as possible, and works well when clusters are compact clouds that are rather well separated from one another. However, when there are large differences in the sizes or geometries of different clusters, as illustrated in Figure 1, the square-error method could split large clusters to minimize the square-error. In the figure, the square-error is larger for the three separate clusters in (a) than for the three clusters in (b) where the big cluster is split into three portions, one of which is merged with the two smaller clusters. The reduction in square-error for (b) is due to the fact that the slight reduction in square error due to splitting the large cluster is weighted by many data points in the large cluster.

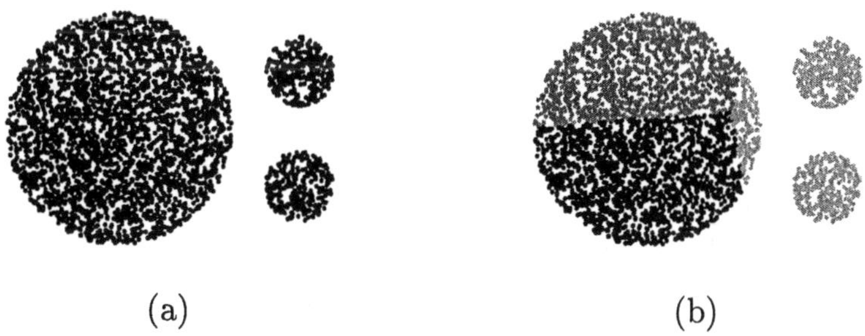

Figure 1: Splitting of a large cluster by partitional algorithms

A hierarchical clustering is a sequence of partitions in which each partition is nested into the next partition in the sequence. An *agglomerative* algorithm for hierarchical clustering starts with the disjoint set of clusters, which places each input data point in an individual cluster. Pairs of items or clusters are then successively merged until the number of clusters reduces to

k. At each step, the pair of clusters merged are the ones between which the distance is the minimum. The widely used measures for distance between clusters are as follows (m_i is the mean for cluster C_i and n_i is the number of points in C_i).

$$\begin{aligned}
d_{mean}(C_i, C_j) &= \|m_i - m_j\| \\
d_{ave}(C_i, C_j) &= 1/(n_i n_j) \sum_{p \in C_i} \sum_{p' \in C_j} \|p - p'\| \\
d_{max}(C_i, C_j) &= \max_{p \in C_i, p' \in C_j} \|p - p'\| \\
d_{min}(C_i, C_j) &= \min_{p \in C_i, p' \in C_j} \|p - p'\|
\end{aligned}$$

For example, with d_{mean} as the distance measure, at each step, the pair of clusters whose centroids or means are the closest are merged. On the other hand, with d_{min}, the pair of clusters merged are the ones containing the closest pair of points. All of the above distance measures have a minimum variance flavor and they usually yield the same results if the clusters are compact and well-separated. However, if the clusters are close to one another (even by outliers), or if their shapes and sizes are not hyperspherical and uniform, the results of clustering can vary quite dramatically. For example, with the data set shown in Figure 1(a), using d_{max}, d_{ave} or d_{mean} as the distance measure results in clusters that are similar to those obtained by the square-error method shown in Figure 1(b). Similarly, consider the example data points in Figure 2. The desired elongated clusters are shown in Figure 2(a). However, d_{mean} as the distance measure, causes the elongated clusters to be split and portions belonging to neighboring elongated clusters to be merged. The resulting clusters are as shown in Figure 2(b). On the other hand, with d_{min} as the distance measure, the resulting clusters are as shown in Figure 2(c). The two elongated clusters that are connected by narrow string of points are merged into a single cluster. This "chaining effect" is a drawback of d_{min} – basically, a few points located so as to form a bridge between the two clusters causes points across the clusters to be grouped into a single elongated cluster.

From the above discussion, it follows that neither the centroid-based approach (that uses d_{mean}) nor the all-points approach (based on d_{min}) work well for non-spherical or arbitrary shaped clusters. A shortcoming of the centroid-based approach is that it considers only one point as representative of a cluster – the cluster centroid. For a large or arbitrary shaped cluster, the centroids of its subclusters can be reasonably far apart, thus causing the

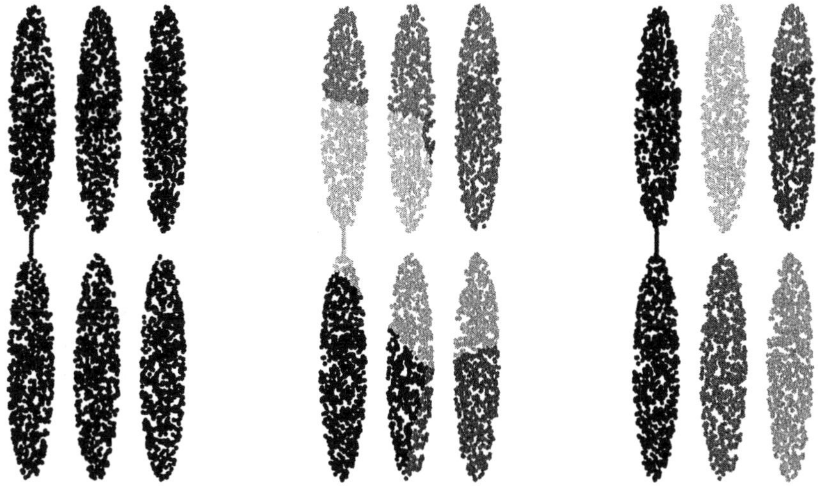

(a) (b) (c)

Figure 2: Clusters generated by hierarchical algorithms

cluster to be split. The all-points approach, on the other hand, considers all the points within a cluster as representative of the cluster. This other extreme, has its own drawbacks, since it makes the clustering algorithm extremely sensitive to outliers and to slight changes in the position of data points.

(a) (b)

Figure 3: Problem of labeling

When the number N of input data points is large, hierarchical clustering algorithms break down due to their non-linear time complexity (typically, $O(N^2)$) and huge I/O costs. In order to remedy this problem, in [76], the authors propose a new clustering method named BIRCH, which represents the state of the art for clustering large data sets. BIRCH first performs a *preclustering phase* in which dense regions of points are represented by compact summaries, and then a centroid-based hierarchical algorithm is used to cluster the set of summaries (which is much smaller than the original dataset).

The preclustering algorithm employed by BIRCH to reduce input size is incremental and approximate. During preclustering, the entire database is scanned, and cluster summaries are stored in memory in a data structure called the CF-tree. For each successive data point, the CF-tree is traversed to find the closest cluster to it in the tree, and if the point is within a threshold distance of the closest cluster, it is absorbed into it. Otherwise, it starts its own cluster in the CF-tree.

Once the clusters are generated, a final labeling phase is carried out in which using the centroids of clusters as seeds, each data point is assigned to the cluster with the closest seed. Using only the centroid of a cluster when redistributing the data in the final phase has problems when clusters do not have uniform sizes and shapes as in Figure 3(a). In this case, as illustrated in Figure 3(b), in the final labeling phase, a number of points in the bigger cluster are labeled as belonging to the smaller cluster since they are closer to the centroid of the smaller cluster.

1.2 Shortcomings of Traditional Clustering Algorithms: Categorical Attributes

Consider a market basket database containing one transaction per customer, each transaction containing the set of items purchased by the customer. The transaction data can be used to cluster the customers such that customers with similar buying patterns are in a single cluster. For example, one cluster may consist of predominantly married customers with infants who buy diapers, baby food, toys etc. (in addition to necessities like milk, sugar and butter), while another may consist of high-income customers that buy imported products like French and Italian wine, Swiss cheese and Belgian chocolate. The clusters can then be used to characterize the different customer groups, and these characterizations can be used in targeted marketing and advertising such that specific products are directed towards specific customer groups. The characterizations can also be used to predict buying patterns of new customers based on their profiles. For example, it may be possible to conclude that high-income customers buy imported foods, and then mail customized catalogs for imported foods to only these high-income customers.

The above market basket database containing transactions is actually an example of a scenario in which attributes of data points are non-numeric. Transactions in the database can be viewed as records with *boolean* attributes, each attribute corresponding to a single item. Further, in the record for a transaction, the attribute corresponding to an item is True if and only if the transaction contains the item; otherwise, it is False. Boolean attributes themselves are a special case of *categorical* attributes. The domain of categorical attributes is not limited to simply True and False values, but could be any arbitrary finite set of values. An example of a categorical attribute is color whose domain includes values such as brown, black, white, etc.

In a typical market basket database, the number of items, and thus the number of attributes is very large (a few thousand) while the size of an average transaction is much smaller (less than a hundred). Furthermore, customers with similar buying patterns and belonging to a single cluster, may buy a small subset of items from a much larger set that defines the cluster. For instance, consider the cluster defined by the set of imported items like French wine, Swiss cheese, Italian pasta sauce, Belgian beer etc. Every transaction in the cluster does not contain all of the above items, but some subset of them. Thus, it is quite possible that a pair of transactions in a cluster have few items in common, but are *linked* by a number of other transactions in the cluster, that have substantial items in common with the two transactions.

The above situation is further exacerbated by the fact that the set of items that define clusters may not have uniform sizes. A cluster involving all the common items such as diapers, baby food and toys will typically involve a large number of items and customer transactions, while the cluster defined by imported products will be much smaller. In the larger cluster, since transactions are spread out over a larger number of items, most transaction pairs will have few items in common and consequently, a smaller percentage of transaction pairs will have a sizable number of items in common. Thus, distances of transactions from the mean in the larger cluster will be much higher. Since the criterion function is defined in terms of distance from the mean, splitting the larger cluster reduces its value, and thus minimizing the criterion function favors splitting large clusters. However, this is not desirable since the large cluster is split even though transactions in the cluster are well connected and strongly linked.

Hierarchical clustering algorithms, too, may be unsuitable for clustering data sets containing categorical attributes. For instance, consider the centroid-based *agglomerative* hierarchical clustering algorithm [25, 46]. In this algorithm, initially, each point is treated as a separate cluster. Pairs of clusters whose centroids or means are the closest are then successively merged until the desired number of clusters remain. For categorical attributes, however, distances between centroids of clusters is a poor estimate of the similarity between them as is illustrated by the following example.

Example 1.1 *Consider a market basket database containing the following 4 transactions over items* 1, 2, 3, 4, 5 *and* 6 *– (a)* {1, 2, 3, 5}, *(b)* {2, 3, 4, 5}, *(c)* {1, 4}, *and (d)* {6}. *The transactions can be viewed as points with boolean (0/1) attributes corresponding to the items* 1, 2, 3, 4, 5 *and* 6. *The four points thus become (1,1,1,0,1,0), (0,1,1,1,1,0), (1,0,0,1,0,0) and (0,0,0,0,0,1). Using euclidean distance to measure the closeness between points/clusters, the distance between the first two points is* $\sqrt{2}$, *which is the smallest distance between pairs of points. As a result, they are merged by the centroid-based hierarchical algorithm. The centroid of the new merged cluster is (0.5,1,1,0.5,1,0). In the next step, the third and fourth points are merged since the distance between them is* $\sqrt{3}$ *which is less than the distance between the centroid of the merged cluster from each of them –* $\sqrt{3.5}$ *and* $\sqrt{4.5}$, *respectively. However, this corresponds to merging transactions* {1, 4} *and* {6} *that don't have a single item in common. Thus, using distances between the centroids of clusters when making decisions about the clusters to merge could cause points belonging to different clusters to be assigned to a single cluster.* □

Once points belonging to different clusters are merged, the situation gets progressively worse as the clustering progresses. What typically happens is a ripple effect – as the cluster size grows, the number of attributes appearing in the mean go up, and their value in the mean decreases. This makes it very difficult to distinguish the difference between two points that differ on few attributes, or two points that differ on every attribute by small amounts. An example will make this issue very clear. Consider the means of two clusters $(\frac{1}{3}, \frac{1}{3}, \frac{1}{3}, 0, 0, 0)$ and $(0, 0, 0, \frac{1}{3}, \frac{1}{3}, \frac{1}{3})$, with roughly the same number of points. Even though, the two clusters have no attributes in common, the euclidean distance between the two means is less than the distance of the point $(1, 1, 1, 0, 0, 0)$ to the mean of the first cluster. Obviously, this is undesirable since the point shares common attributes with the first cluster. An oblivious method based on distance will merge the two clusters and will generate a new cluster with mean $(\frac{1}{6}, \frac{1}{6}, \frac{1}{6}, \frac{1}{6}, \frac{1}{6}, \frac{1}{6})$. Interestingly, the distance of the point $(1,1,1,0,0,0)$ to the new cluster is even larger than the original distance of the point to the first of the merged clusters. In effect, what is happening is that the center of the cluster is spreading over more and more attributes. As this tendency starts, it now becomes closer to other centers which also span a large number of attributes. Thus, these centers tend to spread out in all the attribute values and lose the information about the points in the cluster that they represent. This is why a small ripple soon spreads out to fill all the attributes. This is exactly the behavior that we observed when we ran the centroid-based hierarchical algorithm on real-life data sets [37].

Set theoretic similarity measures such as the *Jaccard coefficient*[1] [46] have often been used, instead of euclidean distance, for document clustering. With the Jaccard coefficient as the distance measure between clusters, centroid-based hierarchical clustering schemes cannot be used since the similarity measure is non-metric, and defined for only points in the cluster and not for its centroid. Thus, we have to use either the *minimum spanning tree* (MST) hierarchical clustering algorithm or hierarchical clustering with *group average* [46]. The MST algorithm merges, at each step, the pair of clusters containing the most similar pair of points while the group average algorithm merges the ones for which the average similarity between pairs of points in the clusters is the highest. The MST algorithm is known to be very sensitive to outliers while the group average algorithm has a tendency to split large clusters (since, as mentioned earlier, the average similarity between two subclusters of a large cluster is small). Furthermore, the Jaccard coefficient is a measure of the similarity between only the two points in question – it thus, does not reflect the properties of the neighborhood of the points. Consequently, the Jaccard coefficient fails to capture the natural clustering of "not so well-separated" data sets with categorical attributes and this is illustrated further in the following example.

Example 1.2 *Consider a market basket database over items* $1, 2, \ldots, 8, 9$. *Consider the 2 transaction clusters shown in Figure 4. The first cluster is defined by 5 items while the second cluster is defined by 4 items. These items are shown at the top of each of the two clusters. Note that items* 1 *and* 2 *are common to both clusters. Each cluster contains transactions of size 3, one for every subset (of size 3) of the set of items that define the cluster.*

[1]The Jaccard coefficient for similarity between transactions T_1 and T_2 is $\frac{|T_1 \cap T_2|}{|T_1 \cup T_2|}$.

<1, 2, 3, 4, 5>		<1, 2, 6, 7>
{1, 2, 3}	{1, 4, 5}	{1, 2, 6}
{1, 2, 4}	{2, 3, 4}	{1, 2, 7}
{1, 2, 5}	{2, 3, 5}	{1, 6, 7}
{1, 3, 4}	{2, 4, 5}	{2, 6, 7}
{1, 3, 5}	{3, 4, 5}	

Figure 4: Basket data example for Jaccard coefficient

The Jaccard coefficient between an arbitrary pair of transactions belonging to the first cluster ranges from 0.2 (e.g., $\{1,2,3\}$ *and* $\{3,4,5\}$*) to 0.5 (e.g.,* $\{1,2,3\}$ *and* $\{1,2,4\}$*). Note that even though* $\{1,2,3\}$ *and* $\{1,2,7\}$ *share common items and have a high Jaccard coefficient of 0.5, they belong to different clusters. In contrast,* $\{1,2,3\}$ *and* $\{3,4,5\}$ *have a lower Jaccard coefficient of 0.2, but belong to the same cluster.*

The MST algorithm may first merge transactions $\{1,2,3\}$ *and* $\{1,2,7\}$ *since the Jaccard coefficient for them has the maximum value of 0.5. Once this happens, the cluster may subsequently merge with transactions from both clusters like* $\{1,3,4\}$ *and* $\{1,6,7\}$ *since these are very similar to transactions in the merged cluster. This is not surprising since the MST algorithm is known to be fragile when clusters are not well-separated.*

The use of group average for merging clusters ameliorates some of the problems with the MST algorithm. However, it may still fail to discover the correct clusters. For instance, similar to MST, it may first merge a pair of transactions containing items 1 *and* 2*, and belonging to different clusters. Note that the group average of the Jaccard coefficient between the new cluster and every other transaction containing both* 1 *and* 2 *is still maximum, that is, 0.5. Consequently, every transaction containing both* 1 *and* 2 *may get merged together into a single cluster in subsequent steps. Thus, in the final clustering, transactions* $\{1,2,3\}$ *and* $\{1,2,7\}$ *from the two different clusters may be assigned to the same cluster.* □

1.3 Two New Directions

In the previous subsections, we showed that traditional clustering algorithms fall short in several interesting situations. In the numeric setting if the clusters are close to one another (even by outliers), or if their shapes and sizes are not hyperspherical and uniform, the results of clustering can vary quite dramatically. This is a significant issue in designing algorithms that are robust and have smooth behaviour.

Furthermore if we are intersted in modeling market baskets or similar behaviour based on categorical attributes, the numeric algorithms perform poorly because of the high dimensional setting. The algorithms also pay in evaluating more expensive distance functions. Alternate ways of computing the distances approximately using similarity estimation introduces issues and errors not necessarily relevant to solving these problems.

In the remainder of this chapter, we focus on two new directions that address the drawbacks of traditional clustering approaches described in the previous subsections. Specifically, we present two robust clustering algorithms, CURE and ROCK, for clustering data in the numeric and categorical settings, respectively.

2 Related Work

In recent years, a number of clustering algorithms for large databases have been proposed [63, 76, 27, 40]. In [63], the authors propose a partitional clustering method for large databases which is based on randomized search. Each cluster is represented by its *medoid*, the most centrally located point in the cluster, and the objective is to find the k best medoids that optimize the criterion function. The authors reduce this problem to that of graph search by representing each set of k medoids as a node in the graph, two nodes being adjacent if they have $k-1$ medoids in common. Initially, an arbitrary node is set to be the current node and a fixed number of iterations are performed. In each iteration, a random neighbor of the current node is set to be the current node if it results in better clustering. The computation of the criterion function for the random neighbor requires the entire database to be examined. It is experimentally shown that CLARANS outperforms the traditional *k-medoid* algorithms. However, CLARANS may require several passes over the database, the runtime cost of which could be prohibitive for large databases. Furthermore, like other partitional clustering algorithms, it could converge to a local optimum.

In [28], the authors use the R*-tree[69, 10, 68] to improve the I/O efficiency of CLARANS on large databases by (1) drawing samples from leaf pages to reduce the number of data points (since data points are packed in leaf nodes based on spatial locality, a sample point in the leaf page can be a good representative point), and (2) focusing on relevant points when evaluating the "goodness" of a neighbor.

Since multiple I/O scans of the data points is a bottleneck for existing clustering algorithms, in [76], the authors present a clustering method named BIRCH whose I/O complexity is a little more than one scan of the data. BIRCH first pre-clusters the data into the maximum possible and finest possible subclusters that can fit in main-memory. For the pre-clustering phase, BIRCH employs a CF-tree which is a balanced tree structure similar to the B-tree and R-tree family[68]. After pre-clustering, BIRCH treats each of the subcluster summaries as representative points, and runs a well-known approximation algorithm from [64], which is an agglomerative hierarchical clustering algorithm.

BIRCH and CLARANS work well for convex or spherical clusters of uniform size. However, they are unsuitable when clusters have different sizes (see Figure 1), or when clusters are non-spherical (see Figure 2). For clustering such arbitrary shaped collections of points (e.g., ellipsoid, spiral, cylindrical), a density-based algorithm called DBSCAN was proposed in [27]. DBSCAN requires the user to specify two parameters that are used to define minimum density for clustering – the radius *Eps* of the neighborhood of a point and the minimum number of points *MinPts* in the neighborhood. Clusters are then found by starting from an arbitrary point and if its neighborhood satisfies the minimum density, including the points in its neighborhood into the cluster. The process is then repeated for the newly added points.

While DBSCAN can find clusters with arbitrary shapes, it suffers from a number of problems. DBSCAN is very sensitive to the parameters *Eps* and *MinPts*, which in turn, are difficult to determine. Furthermore, DBSCAN also suffers from the robustness problems that plague the all-points hierarchical clustering algorithm – in case there is a dense string of points connecting two clusters, DBSCAN could end up merging the two clusters. Also, DBSCAN does not perform any sort of preclustering and executes directly on the entire database. As a result, for large databases, DBSCAN could incur substantial I/O costs. Finally, with density-based algorithms, using random sampling to reduce the input size may not be feasible – the reason for this is that unless sample sizes are large, there could be substantial variations in the density of points within each cluster in the random sample.

In [40], the authors address the problem of clustering related customer transactions in a market basket database. Frequent itemsets used to generate association rules are used to construct a weighted hypergraph. Each frequent itemset is a hyperedge in the weighted hypergraph and the weight of the hyperedge is computed as the average of the confidences for all possible association rules that can be generated from the itemset. Then, a hypergraph partitioning algorithm from [49] is used to partition the items such that the sum of the weights of hyperedges that are cut due to the partitioning is minimized. The result is a clustering of items (not transactions) that occur together in the transactions. Finally, the item clusters are used as the description of the cluster and a scoring metric is used to assign customer transactions to the best item cluster. For example, a transaction T may be assigned to the item cluster C_i for which the ratio $\frac{|T \cap C_i|}{|C_i|}$ is the highest.

The rationale for using item clusters to cluster transactions is questionable. For example, the approach in [40] makes the assumption that itemsets that define clusters are disjoint and have no overlap among them. This may not be true in practice since transactions in different clusters may have a few common items. For instance, consider the market basket database in Example 1.2. With minimum support set to 2 transactions, the hypergraph partitioning algorithm generates two item clusters of which one is $\{7\}$ and the other contains the remaining items (since 7 has the least hyperedges to other items). However, this results in transactions $\{1, 2, 6\}$ and $\{3, 4, 5\}$ being assigned to the same cluster since both have the highest score with respect to the big item cluster.

In [32], for clustering categorical data, Gibson et. al. introduce STIRR, an iterative algorithm based on non-linear dynamic systems. STIRR represents each attribute value as a weighted vertex in a graph, and starting with an initial set of weights on all vertices, "iterates" until a fixed point is reached. Gibson et. al. argue that when the fixed point is reached, the weights naturally partition the set of attribute values for each attribute into two groups: the first with large positive weights and the second with small negative weights. These groups correspond intuitively to projections of clusters on the attribute, and can be used to identify the set of clusters. The CACTUS algorithm, proposed in [31], uses a different summary-based approach for discovering well-defined clusters for categorical data. CACTUS maintains the following two types of summary information: (1) *inter-attribute* summaries consisting of all strongly correlated attribute value pairs from different attributes, and (2) *intra-attribute* summaries consisting of similarities between attribute values of the same attribute. The authors show how the summaries can be constructed efficiently (in a single pass of the dataset), and also propose an algorithm that uses the summary information to discover a set of candidate clusters.

3 CURE: A Tunable Hierarchical Clustering Algorithm for Numeric Attributes

In this section, we describe CURE, a hierarchical clustering algorithm whose salient features are: (1) the clustering algorithm can recognize arbitrarily shaped clusters (e.g., ellipsoidal), (2) the algorithm is robust to the presence of outliers, and (3) the algorithm has linear storage requirements and time complexity of $O(n^2)$ for low-dimensional data. The n data points input to the algorithm are either a sample drawn randomly from the original data points, or a subset of it if partitioning is employed. An analysis of issues related to the size of the random sample and number of partitions is presented in Section 5.

3.1 Intuition and Overview

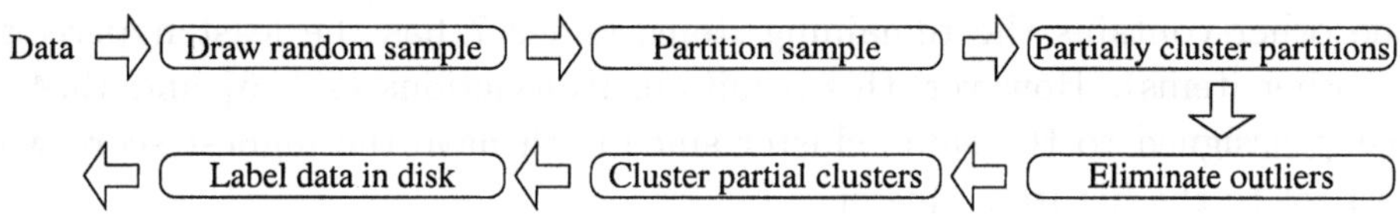

Figure 5: Overview of CURE

The steps involved in clustering using CURE are depicted in Figure 5. The clustering algorithm starts with each input point as a separate cluster, and at each successive step merges the closest pair of clusters. In order to compute the distance between a pair of clusters, for each cluster, c representative points are stored. These are determined by first choosing c *well scattered* points within the cluster, and then shrinking them toward the mean of the cluster by a fraction α. The distance between two clusters is then the distance between the closest pair of representative points – one belonging to each of the two clusters. Thus, only the representative points of a cluster are used to compute its distance from other clusters.

The c representative points attempt to capture the physical shape and geometry of the cluster. Furthermore, shrinking the scattered points toward the mean by a factor α gets rid of surface abnormalities and mitigates the effects of outliers. The reason for this is that outliers typically will be further away from the cluster center, and as a result, the shrinking would cause outliers to move more toward the center while the remaining representative points would experience minimal shifts. The larger movements in the outliers would thus reduce their ability to cause the wrong clusters to be merged. The parameter α can also be used to control the shapes of clusters. A smaller value of α shrinks the scattered points very little and thus favors elongated clusters. On the other hand, with larger values of α, the scattered points get located closer to the mean, and clusters tend to be more compact.

3.2 Clustering Algorithm

In this subsection, we describe the details of our clustering algorithm (see Figure 6). The input parameters to our algorithm are the input data set S containing n points in d-dimensional space and the desired number of clusters k. As we mentioned earlier, starting with the individual points as individual clusters, at each step the closest pair of clusters is merged to form a new cluster. The process is repeated until there are only k remaining clusters.

Data Structures: With every cluster is stored all the points in the cluster. Also, for each cluster u, u.mean and u.rep store the mean of the points in the cluster and the set of c representative points for the cluster, respectively. For a pair of points p, q, $dist(p, q)$ denotes the distance between the points. This distance could be any of the L_p metrics like the L_1 ("manhattan") or L_2 ("euclidean") metrics. Alternatively, nonmetric *similarity functions* can also be used. The distance between two clusters u and v can then be defined as

$$dist(u, v) = \min_{p \in u.\text{rep}, q \in v.\text{rep}} dist(p, q)$$

For every cluster u, we keep track of the cluster closest to it in u.closest.

The algorithm makes extensive use of two data structures – a heap[21] and a k-d tree[68]. Furthermore, corresponding to every cluster, there exists a single entry in the heap – the entries for the various clusters u are arranged in the heap in the increasing order of the distances between u and u.closest. The second data structure is a k-d tree that stores the representative points for every cluster. The k-d tree is a data structure for efficiently storing and retrieving multi-dimensional point data. It is a binary search tree with the distinction that a different key value is tested at each level of the tree to determine the branch to traverse further. For example, for two dimensional data points, the first dimension of the point is tested at even levels (assuming the root node is level 0) while the second dimension is tested at odd levels. When a pair of clusters is merged, the *k-d* tree is used to compute the closest cluster for clusters that may previously have had one of the merged clusters as the closest cluster.

```
procedure cluster(S, k)
begin
1.  T := build_kd_tree(S)
2.  Q := build_heap(S)
3.  while size(Q) > k do {
4.     u := extract_min(Q)
5.     v := u.closest
6.     delete(Q, v)
7.     w := merge(u, v)
8.     delete_rep(T, u); delete_rep(T, v); insert_rep(T, w)
9.     w.closest := x /* x is an arbitrary cluster in Q */
10.    for each x ∈ Q do {
11.       if dist(w, x) < dist(w, w.closest)
12.          w.closest := x
13.       if x.closest is either u or v {
14.          if dist(x, x.closest) < dist(x, w)
15.             x.closest := closest_cluster(T, x, dist(x, w))
16.          else
17.             x.closest := w
18.          relocate(Q, x)
19.       }
20.       else if dist(x, x.closest) > dist(x, w) {
21.          x.closest := w
22.          relocate(Q, x)
23.       }
24.    }
25.    insert(Q, w)
26. }
end
```

Figure 6: Clustering algorithm

Clustering procedure: Initially, for each cluster u, the set of representative points u.rep contains only the point in the cluster. Thus, in Step 1, all

```
procedure merge(u, v)
begin
  w := u ∪ v
  w.mean := (|u|u.mean+|v|v.mean) / (|u|+|v|)
  tmpSet := ∅
  for i := 1 to c do {
     maxDist := 0
     foreach point p in cluster w do {
        if i = 1
           minDist := dist(p, w.mean)
        else
           minDist := min{dist(p, q) : q ∈ tmpSet}
        if (minDist ≥ maxDist){
           maxDist := minDist
           maxPoint := p
        }
     }
     tmpSet := tmpSet ∪ {maxPoint}
  }
  foreach point p in tmpSet do
     w.rep := w.rep ∪ {p + α*(w.mean-p) }
  return w
end
```

Figure 7: Procedure for merging clusters

input data points are inserted into the *k-d* tree. The procedure build_heap (in Step 2) treats each input point as a separate cluster, computes u.closest for each cluster u and then inserts each cluster into the heap (note that the clusters are arranged in the increasing order of distances between u and $u.closest$).

Once the heap Q and tree T are initialized, in each iteration of the while-loop, until only k clusters remain, the closest pair of clusters is merged. The cluster u at the top of the heap Q is the cluster for which u and $u.closest$ are the closest pair of clusters. Thus, for each step of the while-loop, extract_min (in Step 4) extracts the top element u in Q and also deletes u from Q. The merge procedure (see Figure 7) is then used to merge the closest pair of clusters u and v, and to compute new representative points for the new merged cluster w which are subsequently inserted into T (in Step 8). The points in cluster w is simply the union of the points in the two clusters u and v that were merged. The merge procedure, in the for-loop (steps 4-17), first iteratively selects c well-scattered points. In the first iteration, the point farthest from the mean is chosen as the first scattered point. In each subsequent iteration, a point from the cluster w is chosen that is farthest from the previously chosen scattered points. The points are then shrunk toward the mean by a fraction α in Step 16 of the merge procedure.

For the merged cluster w, since the set of representative points for it could have changed (a new set of representative points is computed for it), we need to compute its distance to every other cluster and set w.closest to the cluster closest to it (see steps 11 and 12 of the cluster procedure). Similarly, for a different cluster x in Q, x.closest may change and x may need to be relocated in Q (depending on the distance between x and x.closest). A brute-force method for determining the closest cluster to x is to compute its distance with every other cluster (including w). However, this would require $O(n)$ steps for each cluster in Q, and could be computationally expensive and inefficient. Instead, we observe that the expensive computation of determining the closest cluster is not required for every cluster x. For the few cases that it is required, we use T to determine this efficiently in $O(\log n)$ steps per case. We can classify the clusters in Q into two groups. The first group of clusters are those who had either u or v as the closest cluster before u and v were merged. The remaining clusters in Q constitute the second group. For a cluster x in the first group, if the distance to w is smaller than its distance to the previously closest cluster (say u), then all we have to do is simply set w to be the closest cluster (see Step 17). The reason for this is that we know that the distance between x and every other cluster is greater than the distance between x and u. The problem arises when the distance between x and w is larger than the distance between x and u. In this case, any of the other clusters could become the new closest cluster to x. The procedure closest_cluster (in Step 15) uses the tree T to determine the closest cluster to cluster x. For every point p in x.rep, T is used to determine the nearest neighbor to p that is not in x.rep. From among the nearest neighbors, the one that is closest to one of x's representative points is determined and the cluster containing it is returned as the closest cluster to x. Since we are not interested in clusters whose distance from x is more than $dist(x, w)$, we pass this as a parameter to closest_cluster which uses it to make the search for nearest neighbors more efficient. Processing a cluster x in the second group is much simpler – x.closest already stores the closest cluster to x from among existing clusters (except w). Thus, if the distance between x and w is smaller than x's distance to it's previously closest cluster, x.closest, then w becomes the closest cluster to x (see Step 21); otherwise, nothing needs to be done. In case x.closest for a cluster x is updated, then since the distance between x and its closest cluster may have changed, x may need to be relocated in the heap Q (see steps 18 and 22).

An improved merge procedure: In the merge procedure, the overhead of choosing representative points for the merged cluster can be reduced as follows. The merge procedure, in the outer for-loop (Step 4), chooses c scattered points from among all the points in the merged cluster w. Instead, suppose we selected the c scattered points for w from the $2c$ scattered points for the two clusters u and v being merged (the original scattered points for clusters u and v can be obtained by unshrinking their representative points by α). Then, since at most $2c$ points, instead of $O(n)$ points, need to be examined every time a scattered point is chosen, the complexity of the merge procedure reduces to $O(1)$. Furthermore, since the scattered points for w are chosen from the original scattered points for clusters u and v, they can be expected to be fairly well spread out.

3.3 Time and Space Complexity

We next examine the time and space complexity of our clustering algorithm for n input data points. In the initialization phase, we build Q and T, each of which takes $O(n \log n)$ time. On the average, deleting and inserting an element from the heap takes $O(\log n)$ time [21]. For the k-d tree, it is shown in [29] that for n points, the average cost of inserting, as well as the expected nearest neighbor search time is $O(\log n)$. Similarly, according to [68], the average cost of deleting a randomly selected node from a randomly built k-d tree is $O(\log n)$. The body of the while-loop is executed $O(n)$ times. In Step 7, the cost of merging the two clusters and computing a new set of representative points for the merged cluster is $O(n)$ (since selecting c scattered points requires at most c passes over points in the merged cluster). Furthermore, in Step 8, deleting and inserting the c representative points for the clusters u, v and w takes, on the average, $O(\log n)$ steps. Of the actions performed in the for-loop, the following two actions have the highest overhead and require $O(\log n)$ steps – (1) probing the k-d tree to determine the closest cluster to x (in Step 15), and (2) relocating a cluster in the heap (in Steps 18 and 22). The first is only executed for clusters x for which either u or v were the closest before they were merged. Assuming

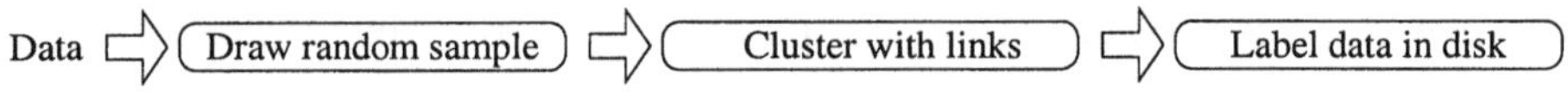

Figure 8: Overview of ROCK

that a cluster can be the closest cluster to at most m clusters, procedure closest_cluster (in Step 15) is invoked in at most $2m$ iterations of the for-loop. The second, updating the heap Q to relocate x, on the other hand, may need to be performed in as many as $3m$ iterations. The only clusters x for which x.closest can change are those for which either u or v were previously closest, or w becomes the new closest cluster. Then, for each iteration of the while loop, we have $O(n)$ iterations of the for-loop, $O(m)$ of which have $O(\log n)$ complexity for finding the closest cluster and relocating clusters in the heap. Therefore, the time complexity of our clustering algorithm is $O(n^2 + nm \log n)$. Since m can be n in worst case, the worst case complexity of our clustering algorithm is $O(n^2 \log n)$.

In [22], the authors show that m is 6 for 2 dimensions, 12 for 3 dimensions and is at most 244 for as many as 8 dimensions. These values are typically much smaller than n, and thus the time complexity of our clustering algorithm is $O(n^2)$ when the dimensionality of data points is small. Since both the heap and the k-d tree require linear space, it follows that the space complexity of our algorithm is $O(n)$.

4 ROCK: A Robust Clustering Algorithm for Categorical Attributes

In this section, we describe the ROCK (RObust Clustering using linKs) algorithm for clustering data with boolean and categorical attributes. We begin by presenting an overview of ROCK, and reserve the details and complexity results for subsequent subsections.

4.1 Overview of ROCK

The steps involved in clustering using ROCK are described in Figure 8. After drawing a random sample from the database, a hierarchical clustering algorithm that employs links is applied to the sampled points. Finally, the clusters involving only the sampled points are used to assign the remaining data points on disk to the appropriate clusters.

In the following subsections, we first present our new clustering model that is based on the notions of *neighbors* and *links*. We also discuss the criterion function that we would like to optimize under our new clustering paradigm. We then describe the steps performed by ROCK in greater detail in Section 4.6.

4.2 Neighbors

Simply put, a point's *neighbors* are those points that are considerably similar to it. Let $sim(p_i, p_j)$ be a *similarity function* that is normalized and captures the closeness between the pair of points p_i and p_j. The function *sim* could be one of the well-known distance metrics (e.g., L_1, L_2) or it could even be *non-metric* (e.g., a distance/similarity function provided by a domain expert). We assume that *sim* assumes values between 0 and 1, with larger values indicating that the points are more similar. Given a threshold θ between 0 and 1, a pair of points p_i, p_j are defined to be *neighbors* if the following holds:

$$sim(p_i, p_j) \geq \theta$$

In the above equation, θ is a user-defined parameter that can be used to control how close a pair of points must be in order to be considered neighbors. Thus, higher values of θ correspond to a higher threshold for the similarity between a pair of points before they are considered neighbors. Assuming that *sim* is 1 for identical points and 0 for totally dissimilar points, a value of 1 for θ constrains a point to be a neighbor to only other identical points. On the other hand, a value of 0 for θ permits any arbitrary pair of points to be neighbors. Depending on the desired closeness, an appropriate value of θ may be chosen by the user.

In the following subsections, we present possible definitions for *sim* for market basket databases and for data sets with categorical attributes.

4.2.1 Market Basket Data

The database consists of a set of transactions, each of which is a set of items. A possible definition based on the Jaccard coefficient [25], for $sim(T_1, T_2)$, the similarity between the two transactions T_1 and T_2, is the following:

$$sim(T_1, T_2) = \frac{|T_1 \cap T_2|}{|T_1 \cup T_2|}$$

where $|T_i|$ is the number of items in T_i. The more items that the two transactions T_1 and T_2 have in common, that is, the larger $|T_1 \cap T_2|$ is, the more similar they are. Dividing by $|T_1 \cup T_2|$ is the scaling factor which ensures that θ is between 0 and 1. Thus, the above equation computes the relative closeness based on the items appearing in both transactions T_1 and T_2.

The above definition of a neighbor rules out subsets of a transaction that are very small in size. A typical example is that of a store where milk is bought by everyone. A transaction with only milk will not be considered very similar to other bigger transactions that contain milk. Also, note that for a pair of transactions T_1 and T_2, *sim* can take at most $\min\{|T_1|, |T_2|\}+1$ values. Thus, there are at most $\min\{|T_1|, |T_2|\} + 1$ distinct similarity levels between the two transactions. As a result, if most transactions have uniform sizes, then there aren't too many possible values for *sim* for the transactions in the database, and this could simplify the choice of an appropriate value for the parameter θ.

4.2.2 Categorical Data

Data sets with categorical attributes can be handled in a manner similar to how we handled market basket data in the previous subsection. Categorical data typically is of fixed dimension and is more structured than market basket data. However, it is still possible that in certain records, values may be *missing* for certain attributes, as is the case for some of the real-life data sets.

We propose to handle categorical attributes with missing values by modeling each record with categorical attributes as a transaction. Corresponding to every attribute A and value v in its domain, we introduce an item $A.v$. A transaction T_i for a record contains $A.v$ if and only if the value of attribute A in the record is v. Note that if the value for an attribute is missing in the record, then the corresponding transaction does not contain items for the attribute. Thus, in the proposal, we simply ignore missing values. The similarity function proposed in the previous subsection can then be used to compute similarities between records by determining the similarity between the corresponding transactions.

Obviously, the above suggested method for dealing with missing values is one of several possible ways to handle them, and may not work well across all domains. For instance, in time-series data, each data point consists of a sequence of time slot, value pairs. We can conceptualize time-series data as a categorical dataset. Each data point can be viewed as a record with every time slot corresponding to a single categorical attribute. The values that are possible in the time slot then constitute the domain of the categorical attribute. Missing values for attributes can frequently result since two individual time-series could be sampled at different times. For example, for young mutual funds that began a year ago, prices for time periods preceding the last year do not exist.

In this case, for two records, in order to compute the similarity between them, we are only interested in considering attributes that have values in both records. This way, if two records are identical for the attributes that do not contain missing values, then we will conclude that the similarity between them is high even though for a number of other attributes, one of the records may have a missing value. Thus, for a pair of records, the transaction for each record only contains items that correspond to attributes for which values are not missing in either record. The similarity between the transactions can then be computed as described earlier in Section 4.2.1. Note that the same record may correspond to different transactions when computing its similarity with respect to different records (depending on the missing values for attributes in the different records).

4.3 Links

Clustering points based on only the closeness or similarity between them is not strong enough to distinguish two "not so well-separated" clusters because it is possible for points in different clusters to be neighbors. In this situation, even if a pair of points p_i and p_j in different clusters are neighbors, it is very unlikely that the pairs have a large number of *common* neighbors, that is, points that are neighbors to both p_i and p_j. This observation motivates the following definition of *links* below that builds on the notion of closeness between points to determine more effectively when close points actually belong to the same cluster.

Let us define $link(p_i, p_j)$ to be the number of *common* neighbors between p_i and p_j. From the definition of links, it follows that if $link(p_i, p_j)$ is large, then it is more probable that p_i and p_j belong to the same cluster. In our framework, we exploit this property of links when making decisions about points to merge into a single cluster. Most existing work only uses the similarity measure between points when clustering them – at each step, points that are the most similar are merged into a single cluster. Since the similarity measure between a pair of points only takes into account characteristics of the points themselves, it is a more *local* approach to clustering. This approach is susceptible to errors since as we mentioned earlier, two distinct clusters may have a few points or outliers that could be very close – relying simply on the similarities between points to make clustering decisions could cause the two clusters to be merged.

The link-based approach adopts a *global* approach to the clustering problem. It captures the global knowledge of neighboring data points into the relationship between individual pairs of points. Thus, since the ROCK clustering algorithm utilizes the information about links between points when making decisions on the points to be merged into a single cluster, it is very robust.

The notion of links between a pair of points, in effect, is the number of distinct paths of length 2 between points p_i and p_j such that every pair of consecutive points on the path are neighbors. Alternative definitions for links, based on paths of length 3 or more, are certainly possible; however, we do not consider these for the following reasons. First and most important, computing paths of length 2 is computationally a lot more efficient than computing paths of higher lengths. Second, points connected by paths of length 2 represent more tightly connected points than points connected by paths with larger lengths. Finally, paths of length 2 constitute the simplest and most cost-efficient way of capturing the knowledge about the mutual neighborhood of points – the additional information gained as a result of considering longer paths may not be as valuable.

Our link-based approach can correctly identify the overlapping clusters in Figure 4. This is because for each transaction, the transaction that it has the most links with is a transaction in its own cluster. For instance, let $\theta = 0.5$ and $sim(T_1, T_2) = \frac{|T_1 \cap T_2|}{|T_1 \cup T_2|}$. Transaction $\{1, 2, 6\}$ has 5 links with transaction $\{1, 2, 7\}$ in its own cluster (due to $\{1, 2, 3\}$, $\{1, 2, 4\}$, $\{1, 2, 5\}$, $\{1, 6, 7\}$ and $\{2, 6, 7\}$) and only 3 links with transaction $\{1, 2, 3\}$ in the other cluster (due to $\{1, 2, 4\}$, $\{1, 2, 5\}$ and $\{1, 2, 7\}$). Similarly, transaction $\{1, 6, 7\}$ has 2 links with every transaction in the smaller cluster (e.g., $\{1, 2, 6\}$) and 0 links with every other transaction in the bigger cluster. Thus, even though the clusters contain common items, with $\theta = 0.5$, our link-based approach would generate the correct clusters shown in Figure 4.

4.4 Criterion Function

For a clustering method, an important question is the following: "is it possible to characterize the best clusters ?". If one could mathematically characterize the "best clusters", then this would aid in the development of algorithms that attempt to find these good clusters. In this subsection, we present a *criterion function* – the best clusters are the ones that maximize the value of the criterion function.

Since we are interested in each cluster to have a high degree of connectivity, we would like to maximize the sum of $link(p_q, p_r)$ for data point pairs p_q, p_r belonging to a single cluster and at the same time, minimize the sum of $link(p_q, p_s)$ for p_q, p_s in different clusters. This leads us to the following criterion function that we would like to maximize for the k clusters.

$$E_l = \sum_{i=1}^{k} n_i * \sum_{p_q, p_r \in C_i} \frac{link(p_q, p_r)}{n_i^{1+2f(\theta)}}$$

where C_i denotes cluster i of size n_i. The rationale for the above criterion function E_l is as follows. It may seem that since one of our goals was to maximize $link(p_q, p_r)$ for all pairs of points p_q, p_r, a simple criterion function like $\sum_{i=1}^{k} \sum_{p_q, p_r \in C_i} link(p_q, p_r)$ that simply sums up the links between pairs of points in the same cluster, ought to work fine. However, even though this criterion function will ensure that points with a large number of links between them are assigned to the same cluster, it does not prevent a clustering in which all points are assigned to a single cluster. Thus, it does not force points with few links between them to be split between different clusters.

In order to remedy the above problem, in the criterion function E_l, we divide the total number of links involving pairs of points in cluster C_i by the expected total number of links in C_i, and then weigh this quantity by n_i, the number of points in C_i. Our estimate for the total number of links in cluster C_i is $n_i^{1+2f(\theta)}$, where $f(\theta)$ is a function that is dependent on the data set as well as the kind of clusters we are interested in, and has the following important property: each point belonging to cluster C_i has approximately $n_i^{f(\theta)}$ neighbors in C_i. If such a function f does exist, then since we can assume that points outside C_i result in a very small number of links to the points in C_i, each point in cluster C_i contributes $n_i^{2f(\theta)}$ links - one for each pair of its neighbors. Thus, we obtain $n_i^{1+2f(\theta)}$ as the expected number of links between pairs of points in C_i. Dividing by the expected number of links in E_l prevents points with very few links between them from being put in the same cluster since assigning them to the same cluster would cause the expected number of links for the cluster to increase more than the actual number of links and the result would be a smaller value for the criterion function.

Of course, it may not be easy to determine an accurate value for function $f(\theta)$. However, we have found that if clusters are fairly well-defined, even an inaccurate but reasonable estimate for $f(\theta)$ can work well in practice [37]. Furthermore, in E_l, every cluster is normalized by $n_i^{1+2f(\theta)}$. Thus, errors in the estimation of $f(\theta)$ affect all the clusters similarly, and does not penalize one cluster excessively over other clusters.

For the market basket data case, one possibility for $f(\theta)$ is $\frac{1-\theta}{1+\theta}$. This can be informally derived under the simplifying assumptions that transactions are of approximately the same size (say t) and are uniformly distributed amongst the (say m) items purchased by customers in cluster C_i. For some constant $c \leq 1$, the number of transactions in the cluster is approximately $\binom{mc}{t}$ and the number of transactions whose similarity to a particular transaction T_i exceeds θ is approximately $\binom{mc}{\frac{(1-\theta)t}{1+\theta}}$ (these are all the transactions that have at least $\frac{2\theta t}{1+\theta}$ items in common with T_i). Thus, the number of neighbors for a transaction in C_i is approximately $n_i^{\frac{1-\theta}{1+\theta}}$ and $f(\theta) = \frac{1-\theta}{1+\theta}$. Intuitively, this makes sense, because when $\theta = 1$, a transaction has only itself as a neighbor and since $f(\theta) = 0$, the expected number of links is in C_i is n_i; on the other hand, when $\theta = 0$, every other transaction in C_i is a neighbor to a transaction and in this case, $f(\theta) = 1$ and the expected number of links in C_i appropriately becomes n_i^3.

In the following subsections, we adapt standard hierarchical clustering so that it attempts to maximize our link-based criterion function.

4.5 Goodness Measure

In Section 4.4, we presented the criterion function which can be used to estimate the "goodness" of clusters. The best clustering of points were those that resulted in the highest values for the criterion function. Since our goal is to find a clustering that maximizes the criterion function, we use a measure similar to the criterion function in order to determine the best pair of clusters to merge at each step of ROCK's hierarchical clustering algorithm. For a pair of clusters C_i, C_j, let $link[C_i, C_j]$ store the number of cross links between clusters C_i and C_j, that is, $\sum_{p_q \in C_i, p_r \in C_j} link(p_q, p_r)$. Then, we define the *goodness measure* $g(C_i, C_j)$ for merging clusters C_i, C_j as follows.

$$g(C_i, C_j) = \frac{link[C_i, C_j]}{(n_i + n_j)^{1+2f(\theta)} - n_i^{1+2f(\theta)} - n_j^{1+2f(\theta)}}$$

The pair of clusters for which the above goodness measure is maximum is the best pair of clusters to be merged at any given step. It seems intuitive that pairs of clusters with a large number of cross links are, in general, good candidates for merging. However, using only the number of cross links between pairs of clusters as an indicator of the goodness of merging them may not be appropriate. This naive approach may work well for well-separated clusters, but in case of outliers or clusters with points that are neighbors, a large cluster may swallow other clusters and thus, points from different clusters may be merged into a single cluster. This is because a large cluster typically would have a larger number of cross links with other clusters.

In order to remedy the problem, as we did in section 4.4, we divide the number of cross links between clusters by the expected number of cross links between them. Thus, if every point in C_i has $n_i^{f(\theta)}$ neighbors, then the expected number of links involving only points in the cluster is approximately $n_i^{1+2f(\theta)}$. Since for large clusters, we can assume that points outside the cluster contribute minimally to the number of links between pairs of points in the cluster, the expected number of links between points within the cluster is approximately $n_i^{1+2f(\theta)}$. As a result, it follows that if two fairly large clusters with sizes n_i and n_j are merged, the number of links between pairs of points in the merged cluster is $(n_i+n_j)^{1+2f(\theta)}$, while the number of links in each of the clusters (before merging) were $n_i^{1+2f(\theta)}$ and $n_j^{1+2f(\theta)}$, respectively. Thus, the expected number of *cross* links or links between pairs of points each from a different cluster, becomes $(n_i+n_j)^{1+2f(\theta)} - n_i^{1+2f(\theta)} - n_j^{1+2f(\theta)}$. We use this normalization factor in the above goodness measure as a heuristic to steer us in the direction of clusters with large values for the criterion function.

4.6 Clustering Algorithm

ROCK's hierarchical clustering algorithm is presented in Figure 9. It accepts as input the set S of n sampled points to be clustered (that are drawn randomly from the original data set), and the number of desired clusters k. The procedure begins by computing the number of links between pairs of points in Step 1 (schemes for this are described in the next subsection). Initially, each point is a separate cluster. For each cluster i, we build a local heap $q[i]$ and maintain the heap during the execution of the algorithm. $q[i]$ contains every cluster j such that $link[i,j]$ is non-zero. The clusters j in $q[i]$ are ordered in the decreasing order of the goodness measure with respect to

```
procedure cluster(S, k)
begin
1.  link := compute_links(S)
2.  for each s ∈ S do
3.      q[s] := build_local_heap(link, s)
4.  Q := build_global_heap(S, q)
5.  while size(Q) > k do {
6.      u := extract_max(Q)
7.      v := max(q[u])
8.      delete(Q, v)
9.      w := merge(u, v)
10.     for each x ∈ q[u] ∪ q[v] do {
11.         link[x, w] := link[x, u] + link[x, v]
12.         delete(q[x], u); delete(q[x], v)
13.         insert(q[x], w, g(x, w)); insert(q[w], x, g(x, w))
14.         update(Q, x, q[x])
15.     }
16.     insert(Q, w, q[w])
17.     deallocate(q[u]); deallocate(q[v])
18. }
end
```

Figure 9: Clustering Algorithm

i, $g(i, j)$.

In addition to the local heaps $q[i]$ for each cluster i, the algorithm also maintains an additional global heap Q that contains all the clusters. Furthermore, the clusters in Q are ordered in the decreasing order of their best goodness measures. Thus, $g(j, \max(q[j]))$ is used to order the various clusters j in Q, where $\max(q[j])$, the max element in $q[j]$, is the best cluster to merge with cluster j. At each step, the max cluster j in Q and the max cluster in $q[j]$ are the best pair of clusters to be merged.

The while-loop in Step 5 iterates until only k clusters remain in the global heap Q. In addition, it also stops clustering if the number of links between every pair of the remaining clusters becomes zero. In each step of the while-loop, the max cluster u is extracted from Q by extract_max and $q[u]$ is used to determine the best cluster v for it. Since clusters u and v will be merged, entries for u and v are no longer required and can be deleted from Q. Clusters u and v are then merged in Step 9 to create a cluster w

containing $|u| + |v|$ points. There are two tasks that need to be carried out once clusters u and v are merged: (1) for every cluster that contains u or v in its local heap, the elements u and v need to be replaced with the new merged cluster w and the local heap needs to be updated, and (2) a new local heap for w needs to be created.

Both these tasks are carried out in the for-loop of Step 10–15. The number of links between clusters x and w is simply the sum of the number of links between x and u, and x and v. This is used to compute $g(x, w)$, the new goodness measure for the pair of clusters x and w, and the two clusters are inserted into each other's local heaps. Note that $q[w]$ can only contain clusters that were previously either in $q[u]$ or $q[v]$ since these are the only clusters that have non-zero links with cluster w. Also, note that, as a result of merging clusters u and v, it is possible that the cluster u or v was previously the best to be merged with x and now w becomes the best one for being merged. Furthermore, it is also possible that neither u nor v was the best cluster to merge with x, but now w is a better cluster to merge with x. For such cases, whenever the max cluster in the local heap for x changes, the algorithm needs to relocate x in Q to reflect information relating to the new best cluster for x (see Step 14). The procedure also needs to make sure that Q contains the best cluster to be merged for the new cluster w.

4.7 Computation of Links

One way of viewing the problem of computing links between every pair of points is to consider an n x n adjacency matrix A in which entry $A[i, j]$ is 1 or 0 depending on whether or not points i and j, respectively, are neighbors. The number of links between a pair of points i and j can be obtained by multiplying row i with column j (that is, $\sum_{l=1}^{n} A[i, l] * A[l, j]$). Thus, the problem of computing the number of links for all pairs of points is simply that of multiplying the adjacency matrix A with itself, in other words, A x A. The time complexity of the naive algorithm to compute the square of a matrix is $O(n^3)$. However the problem of calculating the square of a matrix is a well studied problem and well-known algorithms such as Strassen's algorithm [21] runs in time $O(N^{2.81})$. The best complexity possible currently is $O(N^{2.37})$ due to the algorithm by Coppersfield and Winograd [24].

We expect that, on an average, the number of neighbors for each point will be small compared to the number of input points n, causing the adjacency matrix A to be sparse. For such sparse matrices, the algorithm in Figure 10 provides a more efficient way of computing links.

```
procedure compute_links(S)
begin
Compute nbrlist[i] for every point i in S
Set link[i, j] to be zero for all i, j
for i := 1 to n do {
    N := nbrlist[i]
    for j := 1 to |N| - 1 do
        for  l := j + 1 to |N| do
            link[N[j], N[l]] := link[N[j], N[l]] + 1
}
end
```

Figure 10: Algorithm for computing links

For every point, after computing a list of its neighbors, the algorithm considers all pairs of its neighbors. For each pair, the point contributes one link. If the process is repeated for every point and the link count is incremented for each pair of neighbors, then at the end, the link counts for all pairs of points will be obtained. If m_i is the size of the neighbor list for point i, then for point i, we have to increase the link count by one in m_i^2 entries. Thus, the complexity of the algorithm is $\sum_i m_i^2$ which is $O(nm_m m_a)$, where m_a and m_m are the average and maximum number of neighbors for a point, respectively. In the worst case, the value of m_m can be n in which case the complexity of the algorithm becomes $O(m_a n^2)$. In practice, we expect m_m to be reasonably close to m_a and thus, for these cases, the complexity of the algorithm reduces to $O(m_a^2 n)$ on average. For market basket data, when transactions are uniformly distributed amongst the attributes, we showed that the expected value for the number of neighbors per point is $n^{f(\theta)}$, where $f(\theta) = \frac{1-\theta}{1+\theta}$. Assuming $\theta = 0.5$, m_a is approximately $n^{\frac{1}{3}}$, which is much smaller than $\sqrt{n}$. This results in a time complexity of $O(n^2)$ for computing the links. Note that the list of neighbors for every point can be computed in $O(n^2)$ time. In our experiments, we found that values of θ larger than 0.5 generally resulted in good clustering. For these larger θ values, the overhead of computing links can be expected to be low in practice.

4.8 Time and Space Complexity

Computation of Links: As shown in the previous section, it is possible to compute links among pairs of points in $O(n^{2.37})$ using standard matrix multiplication techniques, or alternatively in $O(n^2 m_a)$ time for average number of neighbors m_a. The space requirement for the link computation is at most $n(n+1)/2$, when every pair of points are linked. However, in general, not every pair of points will have links between them and we expect the storage requirements to be much smaller. We can shown this to be $O(\min\{nm_m m_a, n^2\})$ where m_m is the maximum number of neighbors for a point. This is because a point i can have links to at most $\min\{n, m_m m_i\}$ other points.

Clustering Algorithm: The time to build each local heap initially is $O(n)$ (a heap for a set of n input clusters can be built in time that is linear in the number of clusters [21]). The global heap also has at most n clusters initially, and can be constructed in $O(n)$ time. We next examine the complexities of the steps in the while-loop which is executed $O(n)$ times. The inner for-loop dominates the complexity of the while-loop. Since the size of each local queue can be n in the worst case, and the new merged cluster w may need to be inserted in $O(n)$ local queues, the time complexity of the for-loop becomes $O(n \log n)$, and that of the while-loop is $O(n^2 \log n)$ in the worst case. Due to the above analysis, ROCK's clustering algorithm, along with computation of neighbor lists and links, has a worst-case time complexity of $O(n^2 + nm_m m_a + n^2 \log n)$.

The space complexity of the algorithm depends on the initial size of the local heaps. The reason for this is that when two clusters are merged, their local heaps are deleted and the size of the new cluster's local heap can be no more than the sum of the sizes of the local heaps of the merged clusters. Since each local heap only contains those clusters to which it has non-zero links, the space complexity of ROCK's clustering algorithm is the same as that of link computation, that is, $O(\min\{n^2, nm_m m_a\})$.

5 Enhancements for Large Data Sets

Most hierarchical clustering algorithms, including the one presented in the previous subsection, cannot be directly applied to large data sets due to their quadratic time complexity with respect to the input size. In this section, we present enhancements and optimizations that enable CURE to handle large data sets. We also address the issue of outliers and propose schemes to eliminate them.

5.1 Random Sampling

In order to handle large data sets, we need an effective mechanism for reducing the size of the input to CURE's clustering algorithm. One approach to achieving this is via *random sampling* – the key idea is to apply CURE's clustering algorithm to a random sample drawn from the data set rather than the entire data set. Typically, the random sample will fit in main-memory and will be much smaller than the original data set. Consequently, significant improvements in execution times for CURE can be realized. Also, random sampling can improve the quality of clustering since it has the desirable effect of filtering outliers.

Efficient algorithms for drawing a sample randomly from data in a file in one pass and using constant space are proposed in [75]. As a result, we do not discuss sampling in any further detail, and assume that we employ one of the well-known algorithms for generating the random sample. Also, our experience has been that generally, the overhead of generating a random sample is very small compared to the time for performing clustering on the sample (the random sampling algorithm typically takes less than two seconds to sample a few thousand points from a file containing hundred thousand or more points).

Of course, one can argue that the reduction in input size due to sampling has an associated cost. Since we do not consider the entire data set, information about certain clusters may be missing in the input. As a result, our clustering algorithms may miss out certain clusters or incorrectly identify certain clusters. Even though random sampling does have this tradeoff between accuracy and efficiency, our experimental results indicate that for most of the data sets that we considered, with moderate sized random samples, we were able to obtain very good clusters. In addition, in the following subsection, we use chernoff bounds to analytically derive values for sample sizes for which the probability of missing clusters is low.

5.1.1 Chernoff Bounds

We are interested in answering the following question: what should the size s of the random sample be so that the probability of missing clusters is low ? One assumption that we will make is that the probability of missing a cluster u is low if the sample contains at least $f|u|$ points from the sample, where $0 \leq f \leq 1$. This is a reasonable assumption to make since clusters will usually be densely packed and a subset of the points in the cluster is all that is required for clustering. Furthermore, the value of f depends on the cluster density as well as the intercluster separation – the more well-separated and the more dense clusters become, the smaller is the fraction of the points from each cluster that we need for clustering.

We will use *chernoff bounds* to determine the sample size s for which the probability that the sample contains fewer that $f|u|$ points belonging to cluster u is less than δ, $0 \leq \delta \leq 1$. Let X_j be a random variable that is 1 if the jth point in the sample belongs to cluster u and 0 otherwise. We can assume that all $X_1, X_2, \ldots, X_s$ are independent 0-1 random variables. Note that $X_1, X_2, \ldots, X_s$ are independent Bernoulli trials such that for $1 \leq j \leq s$, $P(X_j = 1) = \frac{|u|}{N}$, where N is the size of the entire data set. Thus, the number of data points in the sample belonging to cluster u is given by the random variable $X = \sum_{j=1}^{s} X_j$. Also, the expected value of X, $\mu = E[X] = E[\sum_{j=1}^{s} X_j] - \sum_{j=1}^{s} E[X_j] - \frac{s|u|}{N}$.

Chernoff bounds state that for $0 < \epsilon \leq 1$, and independent Poisson trials $X_1, X_2, \ldots, X_s$,

$$P[X < (1-\epsilon)\mu] < e^{\frac{-\mu\epsilon^2}{2}} \qquad [61] \tag{1}$$

Thus, due to chernoff bounds, the expression on the right hand side is an upper bound on the probability that the number of u's points in the sample, X, falls below the expected count, μ, by more than $\epsilon\mu$. For our clustering application, we would like the probability of the number of u's points in the sample, X, falling below $f|u|$ to be no more than δ. Stated otherwise,

$$P[X < f|u|] < \delta \tag{2}$$

Equations (1) and (2) can be solved to derive the following value for the minimum sample size.

Theorem 5.1 *For a cluster u, Equation (2) holds if the sample size s satisfies*

$$s \geq fN + \frac{N}{|u|}\log(\frac{1}{\delta}) + \frac{N}{|u|}\sqrt{(\log(\frac{1}{\delta}))^2 + 2f|u|\log(\frac{1}{\delta})} \tag{3}$$

Proof. Rewriting Equation (2) so as to conform to Equation (1), we get

$$P[X < (1 - (1 - \frac{f|u|}{\mu}))\mu] < \delta$$

Applying the chernoff bound result from Equation (1), the above equation holds if

$$e^{\frac{-\mu(1-\frac{f|u|}{\mu})^2}{2}} \leq \delta$$

Substituting $\mu = \frac{s|u|}{N}$ in the above equation, and solving it for s, we obtain the minimum value of s for which Equation (2) holds. □

Thus, based on the above equation, we conclude that for the sample to contain at least $f|u|$ points belonging to cluster u (with high probability), we need the sample to contain more than a fraction f of the total number of points – which seems intuitive. Also, suppose u_{min} is the smallest cluster that we are interested in, and s_{min} is the result of substituting $|u_{min}|$ for $|u|$ in the right hand side of Equation (3). It is easy to observe that Equation (3) holds for $s = s_{min}$ and all $|u| \geq |u_{min}|$. Thus, with a sample of size s_{min}, we can guarantee that with a high probability, $1 - \delta$, the sample contains at least $f|u|$ points from an arbitrary cluster u. Also, assuming that there are k clusters, with a sample size of s_{min}, the probability of selecting fewer than $f|u|$ points from any one of the clusters u is bounded above by $k\delta$.

Equation (3) seems to suggest that the sample size required increases linearly with the input size N. However, this is misleading since as N increases, so does $|u_{min}|$, while the fraction f of each cluster that we need for clustering decreases. Typically, we are not interested in clusters that are very small relative to the input size since these tend to be statistically insignificant and we consider them to be outliers. Thus, we assume that the smallest cluster is some percentage of the average cluster size; that is, $|u_{min}| = \frac{N}{k\rho}$, where $\rho > 1$. Furthermore, depending on the geometry of the smallest cluster, in order to represent it adequately in the sample, we need a constant number of points, ξ, from it. Thus, $f = \frac{\xi}{|u_{min}|}$, and the sample contains $\frac{\xi|u|}{|u_{min}|}$ points from an arbitrary cluster u. From the above discussion, we can derive the following equation for the minimum sample size.

$$s_{min} = \xi k\rho + k\rho \log(\frac{1}{\delta}) + k\rho\sqrt{(\log(\frac{1}{\delta}))^2 + 2\xi \log(\frac{1}{\delta})} \quad (4)$$

This implies that if we only care about clusters whose size is greater than $\frac{N}{k\rho}$ and require ξ points from the smallest cluster, then we require a sample which is *independent of the original number of data points*. For $k = 100$, $\rho = 2$, $\xi = 10$ and $\delta = 0.001$, a sample size of about 6000 is sufficient for clustering. This is independent of N, thus making CURE very scalable.

In [73], the authors propose schemes for mining association rules on a random sample drawn from the data set. However, they require the count of a frequent itemset in the sample to be greater than $\mu(1-\epsilon)$, with high probability (μ is the expected count of the itemset and ϵ is the tolerable error). On the other hand, for clustering, we require the count of a cluster u's points in the sample to be simply greater than $f|u|$ – thus, for us, the required count is independent of the sample size and depends only on the size of the cluster. In the case of [73], however, the required count increases with both the sample size as well as the frequent itemset size.

5.2 Partitioning for Speedup

As the separation between clusters decreases and as clusters become less densely packed, samples of larger sizes are required to distinguish them. However, as the input size n grows, the computation that needs to be performed by CURE's clustering algorithm could end up being fairly substantial due to the $O(n^2 \log n)$ time complexity. In this subsection, we propose a simple partitioning scheme for speeding up CURE when input sizes become large.

The basic idea is to partition the sample space into p partitions, each of size $\frac{n}{p}$. We then partially cluster each partition until the final number of clusters in each partition reduces to $\frac{n}{pq}$ for some constant $q > 1$. Alternatively, we could stop merging clusters in a partition if the distance between the closest pair of clusters to be merged next increases above a certain threshold. Once we have generated $\frac{n}{pq}$ clusters for each partition, we then run a second clustering pass on the $\frac{n}{q}$ partial clusters for all the partitions (that resulted from the first pass).

The idea of partially clustering each partition achieves a sort of *preclustering* – schemes for which were also proposed in [76]. The preclustering algorithm in [76] is incremental and scans the entire data set. Each successive data point becomes part of the closest existing cluster if it is within some threshold distance τ from it – else, it forms a new cluster. Thus, while [76] applies an incremental and approximate clustering algorithm to all the points, CURE uses it's hierarchical clustering algorithm only on the points in a partition. In some abstract sense, CURE's partitioning scheme behaves like a sieve working in a bottom-up fashion and filtering out individual points in favour of partial clusters that are then processed during the second pass.

The advantage of partitioning the input in the above mentioned fashion is that we can reduce execution times by a factor of approximately $\frac{q-1}{pq} + \frac{1}{q^2}$.

The reason for this is that the complexity of clustering any one partition is $O(\frac{n^2}{p^2}(\frac{q-1}{q})\log\frac{n}{p})$ since the number of points per partition is $\frac{n}{p}$ and the number of merges that need to be performed for the number of clusters to reduce to $\frac{n}{pq}$ is $\frac{n}{p}(\frac{q-1}{q})$. Since there are p such partitions, the complexity of the first pass becomes $O(\frac{n^2}{p}(\frac{q-1}{q})\log\frac{n}{p})$. The time complexity of clustering the $\frac{n}{q}$ clusters in the second pass is $O(\frac{n^2}{q^2}\log\frac{n}{q})$. Thus, the complexity of CURE's partitioning algorithm is $O(\frac{n^2}{p}(\frac{q-1}{q})\log\frac{n}{p}+\frac{n^2}{q^2}\log\frac{n}{q})$, which corresponds to an improvement factor of approximately $\frac{q-1}{pq}+\frac{1}{q^2}$ over clustering without partitioning.

An important point to note is that, in the first pass, the closest points in each partition are merged only until the final number of clusters reduces to $\frac{n}{pq}$. By ensuring that $\frac{n}{pq}$ is sufficiently large compared to the number of desired clusters, k, we can ensure that even though each partition contains fewer points from each cluster, the closest points merged in each partition generally belong to the same cluster and do not span clusters. Thus, we can ensure that partitioning does not adversely impact clustering quality. Consequently, the best values for p and q are those that maximize the improvement factor $\frac{q-1}{pq}+\frac{1}{q^2}$ while ensuring that $\frac{n}{pq}$ is at least 2 or 3 times k.

The partitioning scheme can also be employed to ensure that the input set to the clustering algorithm is always in main-memory even though the random sample itself may not fit in memory. If the partition size is chosen to be smaller than the main-memory size, then the input points for clustering during the first pass are always main-memory resident. The problem is with the second pass since the size of the input is the size of the random sample itself. The reason for this is that for every cluster input to the second clustering pass, we store all the points in the cluster. Clearly, this is unnecessary, since our clustering algorithm only relies on the representative points for each cluster. Furthermore, the improved merge procedure in Section 3 only uses representative points of the previous clusters when computing the new representative points for the merged cluster. Thus, by storing only the representative points for each cluster input to the second pass, we can reduce the input size for the second clustering pass and ensure that it fits in main-memory.

5.3 Labeling Data on Disk

Since the input to CURE's clustering algorithm is a set of randomly sampled points from the original data set, the final k clusters involve only a subset of the entire set of points. In CURE, the algorithm for assigning the appropriate cluster labels to the remaining data points employs a fraction of *randomly* selected representative points for each of the final k clusters. Each data point is assigned to the cluster containing the representative point closest to it.

Note that approximating every cluster with multiple points instead a single centroid as is done in [76], enables CURE to, in the final phase, correctly distribute the data points when clusters are non-spherical or non-uniform. The final labeling phase of [76], since it employs only the centroids of the clusters for partitioning the remaining points, has a tendency to split clusters when they have non-spherical shapes or non-uniform sizes (since the space defined by a single centroid is a sphere).

5.4 Handling Outliers

Any data set almost always contains *outliers*. These do not belong to any of the clusters and are typically defined to be points of non agglomerative behaviour. That is, the neighbourhoods of outliers are generally sparse compared to points in clusters, and the distance of an outlier to the nearest cluster is comparatively higher than the distances among points in bonafide clusters themselves.

Every clustering method needs mechanisms to eliminate outliers. In CURE, outliers are dealt with at multiple steps. First, random sampling filters out a majority of the outliers. Furthermore, the few outliers that actually make it into the random sample are distributed all over the sample space. Thus, random sampling further isolates outliers.

In agglomerative hierarchical clustering, initially each point is a separate cluster. Clustering then proceeds by merging closest points first. What this suggests is that outliers, due to their larger distances from other points, tend to merge with other points less and typically grow at a much slower rate than actual clusters. Thus, the number of points in a collection of outliers is typically much less than the number in a cluster.

This leads us to a scheme of outlier elimination that proceeds in two phases. In the first phase, the clusters which are growing very slowly are identified and eliminated as outliers. This is achieved by proceeding with the clustering for some time until the number of clusters decreases below a certain fraction of the initial number of clusters. At this time, we classify clusters with very few points (e.g., 1 or 2) as outliers. The choice of a value for the fraction of initial clusters at which outlier elimination gets triggered is important. A very high value for the fraction could result in a number of cluster points being incorrectly eliminated – on the other hand, with an extremely low value, outliers may get merged into proper clusters before the elimination can kick in. An appropriate value for the fraction, obviously, is dependent on the data set. For most data sets we considered, a value of around $\frac{1}{3}$ performed well.

The first phase of outlier elimination is rendered ineffective if a number of outliers get sampled in close vicinity – this is possible in a randomized algorithm albeit with low probability. In this case, the outliers merge together preventing their elimination, and we require a second level of pruning. The second phase occurs toward the end. Usually, the last few steps of a clustering are the most important ones, because the granularity of the clusters is very high, and a single mistake could have grave consequences. Consequently, the second phase of outlier elimination is necessary for good clustering. From our earlier discussion, it is easy to observe that outliers form very small clusters. As a result, we can easily identify such small groups and eliminate them when there are very few clusters remaining, typically in the order of k, the actual number of clusters desired. We have found the above two-phase approach to outlier elimination to work very well in practice.

6 Clustering from an Optimization Perspective

Clustering has been a benchmark problem in several ways. It would be very difficult to provide a short description of the work in this area. Typically the data is conceived as points in some abstract space and clustering represents the data by a distribution on a small number of regions. The distribution within a region specifies the objective function. The modeling of this distribution, or what constitutes a cluster, leads to a rich definition of clustering. Most natural measures of this clustered property turn out to be NP hard. Faced with this the only analysis tool is to design approximation algorithms that have bounded performance guarantee, that is return a solution which is some factor times the optimum solution.

The K-center and K-median versions of clustering are the most popular models of clustering since the clustering can be specified by providing the k

points. In both versions K points of the data set are chosen as cluster centers; the points are assigned to their nearest cluster center. In K-center the measure of the clustering goodness is the farthest distance of a point from its corresponding center. The measure is reasonable for noise free distributions where the clusters are far apart, but the objective function is sentitive to outliers since the distance from an outlying point to any closeby point can be huge. The K-median problem tries to rectify the problem by minimizing the sum of distances of the points to their respective clusters. Allowing bicriterion optimization in relaxing the description define the Facility Location problem, one of the classical Location Theory problems.

A rich literature exists for the K-center, K-median and facility location problems in exploring heuristics [46, 57, 56], and provable algorithms [41, 50, 55, 54, 71, 34, 17, 15, 7, 52, 48, 14, 59, 3, 47]. The running time is an important issue in considerations of the algorithm "scaling" to the large sets, [30, 11, 44, 35, 72, 60] consider the running time of provable clustering heuristics. [7, 52, 45, 66] consider special metric spaces.

The K-median problem is also relevant in the context of the well known EM and K-means heuristics. The EM (expectation maximization) heuristic iteratively refines a clustering by trying to maximize the probability that the current clustering corresponds to a model. If the clusters are modeled as Gaussian spheres enforcing a partition then the heuristic becomes the clebrated K-Means algorithm. In the K-means heuristic K centers are chosen and subsequently refined to minimize the sum of squares of the distances from the clusters. Thus in each iteration the K-means heuristic replaces a center by the geometric mean of the cluster corresponding to it. The K-median objective function, considering squares of distances, offers an alternative formulation that maximizes the posterii probability. Although the K-median problem is defined on discrete metric spaces, the definition easily extends to real normed spaces. A more detailed discussion of K-means and K-median based algorithms are presented in [39, 65].

A short description of clustering is best suited to context of massive data sets and streaming. For the K-center model of clustering [13] gave a constant factor single pass algorithm $O(nk \log k)$ time and $O(k)$ space algorithm. For K-median, we gave a constant factor single pass approximation in time $\tilde{O}(nk)$ and sublinear n^α space for constant $\alpha > 0$ [35]. Notice that in a single pass the data points cannot be labeled with the clusters they belong to. These algorithms output the cluster centers only. A more detailed description of streaming algorithms and clustering is in [39].

An interesting and important direction is also considering clustering which is accurate for most data points [19]. More results in this direction are presented in [18]. Another important context in which clustering has been considered is the property testing context [58, 2]. In property testing models the goal is to estimate a property, in the current discussion if the clustering objective function is below a certain value, by considering a tiny fraction of the input. The property tester answers positive if the condition is true and false if the condition is far from true, that is the best possible objective function value is significantly larger than the queried value. Not surprisingly, the property testing frame of algorithms rely on sampling and provide guarantees relating the scaled solution to the sample and the solution to the entire data.

The K-center and K-median are one of many models extant in theory literature. In both the clusters are modeled as stars, each point is assigned to a cluster "center" which is the hub. These objects can be measured in very different ways, K-center is concerned with the maximum radius and K-median the pointwise average radius. Variations are known measuring maximum diameter [38], sum of cluster radii [23], and several others [16]. Alternately it is possible to model the clusters as cliques and then the objective function can be thought of minimizing sum of pairwise distances of points belonging to the same cluster (minsum clustering) [9] or maximizing the sum of pairwise distances of points not in the same cluster (k-cut) [74]. The clustering can also be modeled as spanning trees and this results in heuristics reliant on the minimum spanning tree or forest. The duality of modeling clusters as agglomerative, as defined by connected structures, or partitional as defined by cuts results in very different objective functions which require significantly differing techniques in developing bounded performance heuristics.

Formulations independent of distances exist as well. [53] considered clustering from graph conductance perspectives, loosely speaking the clusters define regions in the graph from which it is difficult to "escape" if we were performing a random walk. [26, 67, 6] consider clustering defined by projections onto subspaces.

The above is hardly an exhaustive summary, more appropriately is a discussion of the basic models which have been analyzed. Most of these problems have a large number of variants, for example in relation to network design, which we omit discussion in interest of space.

It is also impossible to discuss theoretical clustering algorithms without considering similarity estimation. Similarity estimation is directly relevant to streaming and massive data sets where repeatedly evaluating a complicated distance function is prohibitive. One of the most celebrated problem in this case is dimensionality reduction of real normed spaces. A rich literature exists in this context, building upon [20, 5, 51, 42, 43], which would not be possible to discuss here. Similar questions have been posed on estimating similarity of strings [12, 62], sets [8], trees [77, 33]. As with clustering variants, these are also mere pointers to a much larger body of work.

7 Summary

Traditional clustering algorithms either favor clusters with spherical shapes and similar sizes, or are very fragile in the presence of outliers. Further, clustering algorithms that use distances between points for clustering are not appropriate for boolean and categorical attributes. In this chapter, we presented two new directions for robustly clustering large datasets in the presence of both numerical and categorical attributes.

Our first clustering algorithm called CURE is more robust to outliers, and identifies clusters having non-spherical shapes and wide variances in size. CURE achieves this by representing each cluster by a certain fixed number of points that are generated by selecting *well scattered* points from the cluster and then shrinking them toward the center of the cluster by a specified fraction. Having more than one representative point per cluster allows CURE to adjust well to the geometry of non-spherical shapes and the shrinking helps to dampen the effects of outliers. To handle large databases, CURE employs a combination of *random sampling* and *partitioning*. A random sample drawn from the data set is first partitioned and each partition is *partially* clustered. The partial clusters are then clustered in a second pass to yield the desired clusters.

For data with boolean and categorical attributes, our second clustering algorithm ROCK, employs the novel concept of *links* to measure the similarity/proximity between a pair of data points. In a nutshell, the number of links between a pair of points is the number of common neighbors shared by the points. Unlike distances or similarities between a pair of points which are local properties involving only the two points in question, the link concept incorporates global information about the other points in the neighborhood of the two points. The ROCK hierarchical clustering algorithm employs links and not distances when merging clusters, and is thus more robust since links inject global knowledge into the clustering process.

References

[1] R. Agrawal, J. Gehrke, D. Gunopulos, and P. Raghavan. Automatic subspace clustering of high dimensional data for data mining. *Proceedings of the Symposium on Management of Data (SIGMOD)*, 1998.

[2] N. Alon, S. Dar, M. Parnas, and D. Ron. Testing of clustering. *Proceedings of the Symposium on Foundations of Computer Science (FOCS)*, 2000.

[3] Vijay Arya, Naveen Garg, Rohit Khandekar, Kamesh Munagala, and Vinayaka Pandit. Local search heuristic for k-median and facility location problems. In *Proceedings of the Symposium on Theory of Computing (STOC)*, pages 21–29, 2001.

[4] Rakesh Agrawal, King-Ip Lin, Harpreet S. Sawhney, and Kyuseok Shim. Fast similarity search in the presence of noise, scaling, and translation in time-series databases. *Proceedings of the International Conference on Very Large Databases (VLDB)*, pages 490–501, 1995.

[5] Sunil Arya, David M. Mount, Nathan S. Netanyahu, Ruth Silverman, and Angela Y. Wu. An optimal algorithm for approximate nearest neighbor searching fixed dimensions. *Journal of the ACM*, 45(6):891–923, 1998.

[6] Pankaj K. Agarwal and Cecilia Procopiuc. Approximation algorithms for projective clustering. *Proceedings of the ACM-SIAM Symposium on Discrete Algorithms (SODA)*, pages 538–547, 2000.

[7] Sanjeev Arora, Prabhakar Raghavan, and Satish Rao. Approximation schemes for euclidean k -medians and related problems. In *Proceedings of the Symposium on Theory of Computing (STOC)*, pages 106–113, 1998.

[8] Andrei Z. Broder, Moses Charikar, Alan M. Frieze, and Michael Mitzenmacher. Min-wise independent permutations. *Journal of Computer and System Sciences*, 60(3):630–659, 2000.

[9] Y. Bartal, M. Charikar, and D. Raz. Approximating min-sum k-clustering in metric spaces. *Proceedings of the Symposium on Theory of Computing (STOC)*, 2001.

[10] N. Beckmann, H. Kriegel, R. Schneider, and B. Seeger. The R^*-tree: an efficient and robust access method for points and rectangles. *Proceedings of the ACM SIGMOD Conference on Management of Data*, pages 322–331, 1990.

[11] A. Borodin, R. Ostrovsky, and Y. Rabani. Subquadratic approximation algorithms for clustering problems in high dimensional spaces. *Proceedings of the Symposium on Theory of Computing (STOC)*, 1999.

[12] Andrei Z. Broder. On the resemblance and containment of documents. In *Compression and Complexity of Sequences (SEQUENCES '97)*, pages 21–29. IEEE Computer Society, 1998.

[13] Moses Charikar, Chandra Chekuri, Tomas Feder, and Rajeev Motwani. Incremental clustering and dynamic information retrieval. In *ACM Symposium on Theory of Computing*, pages 626–635, 1997.

[14] Moses Charikar and Sudipto Guha. Improved combinatorial algorithms for the facility location and k-median problems. In *IEEE Symposium on Foundations of Computer Science*, pages 378–388, 1999.

[15] M. Charikar, S. Guha, É. Tardos, and D. B. Shmoys. A constant factor approximation algorithm for the k-median problem. *Proceedings of the Thirty-First Annual ACM Symposium on Theory of Computing*, 1999.

[16] M. Charikar. Approximation algorithms for clustering problems. *PhD Thesis, Stanford University*, 2000.

[17] F. Chudak. Improved approximation algorithms for uncapacitated facility location. *Proceedings of Integer Programming and Combinatorial Optimization*, LNCS 1412:180–194, 1998.

[18] Moses Charikar, Samir Khullera, David M. Mount, and Giri Narasimhan. Algorithms for facility location problems with outliers. *Proceedings of the ACM-SIAM Symposium on Discrete Algorithms (SODA)*, pages 642–651, 2001.

[19] D. Cutting, D. Karger, Jan Pedersen, and J. Tukey. Scatter/gather: A cluster-based approach to browsing large document collections. *SIGIR*, 1992.

[20] K. L. Clarkson. A randomized algorithm for closestpoint queries. *SIAM Journal on Computing*, 17, 1988.

[21] Thomas H. Cormen, Charles E. Leiserson, Ronald L. Rivest, and Clifford Stein. *Introduction to Algorithms, 2nd ed.* MIT Press, 2001.

[22] H. S. M. Coxeter. An upper bound for the number of equal nonoverlapping speheres that can touch each another of the same size. *Symposia in Pure Mathematics*, 7:53–71, 1964.

[23] Moses Charikar and Rina Panigrahy. Clustering to minimize the sum of cluster diameters. *Proceedings of the Symposium on Theory of Computing (STOC)*, pages 1–10, 2001.

[24] C. Coppersmith and S. Winograd. Matrix multiplication via arithmetic progressions. *Proceedings of the Symposium on Theory of Computing (STOC)*, 1987.

[25] R. O. Duda and P. E. Hart. *Pattern Classification and Scene Analysis.* Wiley, New York, 1973.

[26] P. Drineas, R. Kannan, A. Frieze, and V. Vinay. Clustering in large graphs and matrices. *Proceedings of the ACM-SIAM Symposium on Discrete Algorithms (SODA)*, 1999.

[27] M. Ester, H. Kriegel, J. Snader, and X. Xu. A density-based algorithm for discovering clusters in large spatial database with noise. *International Conference on Knowledge Discovery in Databases and Data Mining (KDD-96)*, 1996.

[28] M. Ester, H. Kriegel, and X. Xu. A database interface for clustering in large spatial databases. *International Conference on Knowledge Discovery in Databases and Data Mining (KDD-95)*, 1995.

[29] J. H. Friedman, J. L. Bentley, and R. A. Finkel. An algorithm for finding best matches in logarithmic expected time. *ACM Transactions on Mathematical Software*, 3:209–226, 1977.

[30] Toms Feder and Daniel H. Greene. Optimal algorithms for appropriate clustering. *Proceedings of the Symposium on Theory of Computing (STOC)*, pages 434–444, 1988.

[31] V. Ganti, J. Gehrke, and R. Ramakrishnan. CACTUS–Clustering categorical data using summaries. *International Conference on Knowledge Discovery in Databases and Data Mining (KDD-99)*, 1999.

[32] D. Gibson, J. Kleinberg, and P. Raghavan. Clustering categorical data: An approach based on dynamical systems. *Proceedings of the 24'th International Conference on Very Large Data Bases*, 1998.

[33] S. Guha, H. Jagadish, N. Koudas, D. Srivastava, and T. Yu. Approximate xml joins. *Proceedings of the Symposium on Management of Data (SIGMOD)*, pages 287–298, 2002.

[34] S. Guha and S. Khuller. Greedy strikes back: Improved facility location algorithms. *Proceedings of the 9th Annual ACM-SIAM Symposium on Discrete Algorithms*, pages 649–657, 1998.

[35] S. Guha, N. Mishra, R. Motwani, and L. O'Callaghan. Clustering data streams. *Proceedings of the Symposium on Foundations of Computer Science (FOCS)*, 2000.

[36] S. Guha, R. Rastogi, and K. Shim. CURE: An efficient algorithm for clustering large databases. *Proceedings of the Symposium on Management of Data (SIGMOD)*, 1998.

[37] S. Guha, R. Rastogi, and K. Shim. ROCK: A robust clustering algorithm for categorical attributes. *Proceedings of ICDE*, 1999.

[38] T. F. Gonzalez. Clustering to minimize the maximum intercluster distance. *Theoretical Computer Science*, pages 293–306, 1985.

[39] Sudipto Guha. Approximation algorithms for facility location problems. *Ph.D. Thesis, Stanford University*, 2000.

[40] E. Han, G. Karypis, V. Kumar, and B. Mobasher. Clustering based on association rule hypergraphs. *Technical report, 1997 SIGMOD Workshop on Research Issues on Data Mining and Knowledge Discovery*, 1997.

[41] D. Hochbaum and D. B. Shmoys. A best possible heuristic for the k-center problem. *Math of Operations Research*, 10(2):180–184, 1985.

[42] P. Indyk and R. Motwani. Approximate Nearest Neighbors: Towards Removing the Curse of Dimensionality. *Proceedings of the Symposium on Theory of Computing (STOC)*, 1998.

[43] Piotr Indyk, Rajeev Motwani, Prabhakar Raghavan, and Santosh Vempala. Locality-preserving hashing in multidimensional spaces. *Proceedings of the Symposium on Theory of Computing (STOC)*, pages 618–625, 1997.

[44] P. Indyk. Sublinear time algorithms for metric space problems. *Proceedings of the Symposium on Theory of Computing*, 1999.

[45] Piotr Indyk. A sublinear time approximation scheme for clustering in metric spaces. *Proceedings of the Symposium on Foundations of Computer Science (FOCS)*, pages 154–159, 1999.

[46] A. K. Jain and R. C. Dubes. *Algorithms for Clustering Data.* Prentice Hall, 1988.

[47] K. Jain, M. Mahdian, and A. Saberi. A new greedy approach for facility location problem. *Proceedings of the Symposium on Theory of Computing (STOC)*, 2002.

[48] K. Jain and V. Vazirani. Primal-dual approximation algorithms for metric facility location and k-median problems. *Proceedings of the Twenty-Ninth Annual IEEE Symposium on Foundations of Computer Science*, 1999.

[49] George Karypis, Rajat Aggarwal, Vipin Kumar, and Shashi Shekhar. Multilevel hypergraph partitioning: Application in VLSI domain. *Proceedings of the ACM/IEEE Design Automation Conference*, 1997.

[50] O. Kariv and S. L. Hakimi. An algorithmic approach to network location problems, part ii: p-media ns. *SIAM Journal on Applied Mathematics*, pages 539–560, 1979.

[51] Eyal Kushilevitz, Rafail Ostrovsky, and Yuval Rabani. Efficient search for approximate nearest neighbor in high dimensional spaces. *Proceedings of the Symposium on Theory of Computing (STOC)*, pages 614–623, 1998.

[52] S. Kolliopoulos and S. Rao. A nearly linear-time approximation scheme for the euclidean k-median problem. *Proc. 7th European Symposium on Algorithms*, pages 378–389, 1999.

[53] Ravi Kannan, Santosh Vempala, and Adrian Vetta. On clusterings: Good, bad and spectral. *Proceedings of the Symposium on Foundations of Computer Science (FOCS)*, pages 367–377, 2000.

[54] J. H. Lin and J. S. Vitter. Approximation algorithms for geometric median problems. *Information Processing Letters*, 44:245–249, 1992.

[55] J. H. Lin and J. S. Vitter. ϵ-approximations with minimum packing constraint violations. *Proceedings of the Twenty-Fourth Annual ACM Symposium on Theory of Computing*, 1992.

[56] O. L. Managasarian. Mathematical programming in data mining. *Data Mining and Knowledge Discovery*, 1997.

[57] P. Mirchandani and R. Francis, editors. *Discrete Location Theory.* John Wiley and Sons, Inc., New York, 1990.

[58] Nina Mishra, Dan Oblinger, and Leonard Pitt. Sublinear time approximate clustering. *Proceedings of the ACM-SIAM Symposium on Discrete Algorithms (SODA)*, 2001.

[59] R. Mettu and C. G. Plaxton. The onlike median problem. *Proceedings of the 41st IEEE Foundations of Computer Science*, 2000.

[60] Ramgopal R. Mettu and C. Greg Plaxton. Optimal time bounds for approximate clustering. *Manuscript*, 2002.

[61] R. Motwani and P. Raghavan. *Randomized Algorithms.* Cambridge University Press, 1995.

[62] S. Muthukrishnan. Efficient algorithms for document retrieval problems. *Proceedings of the ACM-SIAM Symposium on Discrete Algorithms (SODA)*, 2002.

[63] Raymond T. Ng and Jiawei Han. Efficient and effective clustering methods for spatial data mining. *Proceedings of the 20'th International Conference on Very Large Data Bases*, 1994.

[64] C. F. Olson. Parallel algorithms for hierarchical clustering. *Technical report, University of California at Berkeley*, 1993.

[65] Liadan O'Callaghan, Nina Mishra, Adam Meyerson, Sudipto Guha, and Rajeev Motwani. Streaming-data algorithms for high-quality clustering. *Proceedings of ICDE*, 2002.

[66] Rafail Ostrovsky and Yuval Rabani. Polynomial time approximation schemes for geometric k-clustering. *Proceedings of the Symposium on Foundations of Computer Science (FOCS)*, 2000.

[67] Cecilia Procopiuc, Michael Jones, Pankaj K. Agarwal, and T. M. Murali. A monte carlo algorithm for fast projective clustering. *Proceedings of the Symposium on Management of Data (SIGMOD)*, 2002.

[68] Hanan Samet. *The Design and Analysis of Spatial Data Structures.* Addison Wesley, 1990.

[69] T. Sellis, N. Roussopoulos, and C. Faloutsos. The R^+ tree: a dynamic index for multi-dimensional objects. *Proceedings of the 13th International Conference on Very Large Data Bases*, pages 507–518, 1987.

[70] Kyuseok Shim, Ramakrishnan Srikant, and Rakesh Agrawal. High-dimensional similarity joins. pages 301–311, 1997.

[71] D. B. Shmoys, É. Tardos, and K. Aardal. Approximation algorithms for facility location problems. *Proceedings of the Twenty-Ninth Annual ACM Symposium on Theory of Computing*, pages 265–274, 1997.

[72] Mikkel Thorup. Quick k-median, k-center, and facility location for sparse graphs. *ICALP*, pages 249–260, 2001.

[73] H. Toivonen. Samping large databases for association rules. *Proceedings of the International Conference on Very Large Databases (VLDB)*, 1996.

[74] Vijay Vazirani. *Approximation Algorithms.* Springer Verlag, 2001.

[75] J. S. Vitter. Random sampling with a reservoir. *ACM Transactions on Mathematical Software*, 11(1):37–57, 1985.

[76] Tian Zhang, Raghu Ramakrishnan, and Miron Livny. Birch: An efficient data clustering method for very large databases. *Proceedings of the ACM SIGMOD Conference on Management of Data*, pages 103–114, 1996.

[77] K. Zhang and D. Sasha. Tree pattern matching. In Apocolisto and Galil, editors, *Pattern Matching Algorithms.* Oxford University Press, 1997.

CLUSTERING AND INFORMATION RETRIEVAL (pp. 83-103)
W. Wu, H. Xiong and S. Shekhar(Eds.)

Finding Topics in Collections of Documents: A Shared Nearest Neighbor Approach [1]

Levent Ertöz
Department of Computer Science and Engineering
University of Minnesota, Minneapolis, MN 55455
E-mail: ertoz@cs.umn.edu

Michael Steinbach
Department of Computer Science and Engineering
University of Minnesota, Minneapolis, MN 55455
E-mail: steinbac@cs.umn.edu

Vipin Kumar
Department of Computer Science and Engineering
University of Minnesota, Minneapolis, MN 55455
E-mail: kumar@cs.umn.edu

Contents

[1]This work was partially supported by NSF grant ACI-9982274, by LLNL/DOE grant #B347714, and by Army High Performance Computing Research Center contract number DAAH04-95-C-0008. The content of this work does not necessarily reflect the position or policy of the government and no official endorsement should be inferred. Access to computing facilities was provided by AHPCRC and the Minnesota Supercomputing Institute.

1 Introduction

Given a set of documents, clustering is often used to group the documents, in the hope that each group will represent documents with a common theme or topic. Initially, hierarchical clustering was used to cluster documents [5]. This approach has the advantage of producing a set of nested document clusters, which can be interpreted as a topic hierarchy or tree, from general to more specific topics. In practice, while the clusters at different levels of the hierarchy sometimes represent documents with consistent topics, it is common for many clusters to be a mixture of topics, even at lower, more refined levels of the hierarchy. More recently, as document collections have grown larger, K-means clustering has emerged as a more efficient approach for producing clusters of documents [4, 9, 16]. K-means clustering produces a set of un-nested clusters, and the top (most frequent or highest "weight") terms of the cluster are used to characterize the topic of the cluster. Once again, it is not unusual for some clusters to be mixtures of topics.

Our goal was to find an algorithm that would more consistently produce clusters of documents with strong, coherent themes, even if it were necessary to omit many of the documents in the process. After all, in an arbitrary collection of documents, e.g., a set of newspaper articles, there is no reason to expect that all documents belong to a group with a strong topic or theme. While this approach does not provide a complete organization of all documents, it does identify the "nuggets" of information in a document collection and can profitably be applied to practical problems such as grouping the search results of a Web search engine.

To accomplish our goal of consistently finding coherent clusters, we developed a model of documents that allows us to clearly define what we mean by a coherent cluster. This model is based on the idea that a group of documents with a strong theme or topic is characterized by its use of words from a small number of specialized vocabularies. For example, documents about the Olympics would, in addition to words from a general vocabulary, tend to have many sports terms and international terms, i.e., country names. This model can be formalized as a generative probabilistic model, where each cluster of documents that represents a coherent concept is generated by selecting words from a few specialized vocabularies and one general vocabulary.

Our model indicates, quite realistically, that any individual document may be more similar to a document in another coherent cluster than to any document in its own cluster [16]. For example, a particular document about the Olympics may be more similar to another sports document, than to any other document about the Olympics. However, clustering methods, such as K-means and hierarchical clustering, tend to assume that an object belongs to a particular cluster only if it is closer to at least some object in that cluster then to some object in other clusters. To overcome this problem, we use a clustering approach based on how many nearest neighbors a document shares [8]. For documents, this somewhat indirect measure of similarity turns out to be more accurate than a direct similarity measure based, say, on the cosine measure. This approach also deals with another problem of document clusters, i.e., that there is a lot of variation in the "tightness" of different clusters. In such cases, many clustering techniques, e.g., K-means, will either combine two tight clusters into one cluster or split a loose cluster.

The basic outline of this paper is as follows. Section 2 provides a brief background in documents and clustering, and discusses some related work. Section 3 introduces our shared nearest neighbor clustering algorithm, while section 4 describes our document model. Section 5 presents some experimental results comparing our shared nearest neighbor clustering approach to K-means. Section 6 is a brief conclusion and an indication of areas for future work.

2 Document and Clustering Preliminaries

Documents are represented using the vector-space model [15, 11]. In particular, we remove stop words, perform stemming using Porter's suffix-stripping algorithm, and then weight each term based on its *inverse document frequency* (IDF). Finally, each document vector is normalized to unit length. To compute the similarity between documents, we used the cosine measure [11].

The two most common techniques used for clustering documents are hierarchical and partitional (K-means) clustering techniques [3, 12]. Hierarchical techniques produce a nested sequence of partitions, with a single, all-inclusive cluster at the top and singleton clusters of individual points at the bottom. Each intermediate level can be viewed as combining two clusters from the next lower level (or splitting a cluster from the next higher level). In contrast to hierarchical techniques, partitional clustering techniques create a one-level (un-nested) partitioning of the data points (documents). There are a number of partitional techniques, but we shall only consider the K-means algorithm, which is widely used in document clustering. K-means is based on the idea that a center point can represent a cluster. For K-means, we use the notion of a centroid, which is the mean of a group of points.

2.1 Related Work - General

Here we restrict our focus to the use of clustering for topic or theme related document tasks. Clustering has been proposed for use in browsing a collection of documents [2] or in organizing the results returned by a search engine in response to a user's query [18]. Document clustering has also been used to automatically generate hierarchical clusters of documents [13]. (The automatic generation of a taxonomy of Web documents like that provided by Yahoo! (www.yahoo.com) is often cited as a goal.) A somewhat different approach [1] finds the natural clusters in an already existing document taxonomy (Yahoo!), and then uses these clusters to produce an effective document classifier for new documents. Recent work to generate document hierarchies [14] uses some of the clustering techniques from [2]. Much recent work in document clustering has focused on K-means clustering [4, 9, 16]. Indeed, it has been suggested that variants of the K-means algorithm can produce both nested and un-nested sets of clusters that are as good or better than those produced by traditional hierarchal clustering techniques [16]. For that reason we only compare our algorithm to K-means. For recent developments comparing K-means and hierarchical clustering, we refer the reader to [19].

Finally, note that our work is not closely related to Topic Detection and Tracking (TDT), which has a more temporal flavor. Also, our definition of a topic is not the same as that employed in TDT work. Instead, our work derives from work in document clustering, which has the goal of improved browsing or organization of documents.

2.2 Related Work – Specific

Our clustering algorithm is based on a shared nearest neighbor clustering algorithm described in [8]. A similar approach, but for hierarchical clustering, was developed in [6]. Recently, a couple of other clustering algorithms have used shared nearest neighbor ideas [7, 10]. The work in [7] also gives an example of a situation where a measure of similarity, which is based on the number of neighbors two points share, was used to overcome the problem that two data points might belong to separate classes, but still be most similar to each other.

We explain the approach of [8], which we call Jarvis-Patrick clustering, in more detail in preparation for Section 3. First note that while cluster analysis sometimes uses the original data matrix, many clustering algorithms use a similarity matrix, which is an m by m matrix (m = number of objects) containing all the pairwise similarities between the objects being considered. If x_i and x_j are the ith and j^{th} objects,respectively, then the entry at the ith row and j^{th} column of the similarity matrix is the similarity, s_{ij} between x_i and x_j. A similarity matrix defines a weighted graph, where the nodes are the points being clustered, and the weighted edges represent the similarities between points, i.e., the entries of the similarity matrix.

Thus, from a graph point of view, clustering is equivalent to breaking the graph into connected components, one for each cluster. We will describe the shared nearest neighbor algorithm in [8] in these terms. First the n nearest neighbors of all points are found. In graph terms this can be regarded as breaking all but the n strongest links between a point and all other points in the proximity graph. This forms what we call a "nearest neighbor graph." Note that the nearest neighbor graph is just a sparsified version of the original similarity graph, which is derived by breaking the links to less similar points.

We then determine the number of nearest neighbors shared by any two points. In graph terminology, we form what we call the "shared nearest neighbor" graph. We do this by replacing the weight of each link between two points (in the nearest neighbor graph) by the number of neighbors that the points share. In other words, this is the number of length 2 paths between any two points in the nearest neighbor graph [7].

After, this shared nearest neighbor graph is created, all pairs of points are compared and if any two points share more than T neighbors, i.e., have a link in the shared nearest neighbor graph with a weight more than our threshold value, $T(T \leq n)$, then the two points and any cluster they are part of are merged. In other words, clusters are connected components in our shared nearest neighbor graph after we sparsify using a threshold.

This approach has a number of nice properties. It can handle clusters of different densities since the shared nearest neighbor approach is self-scaling. Also, this approach is transitive, i.e., if point, p, shares lots of nearest neighbors with point, q, which in turn shares lots of nearest neighbors with point, r, then points p, q and r all belong to the same cluster. The transitive property, in turn, allows this technique to handle clusters of different sizes and shapes. We have extended the Jarvis-Patrick approach as described in the next section.

3 Our Clustering Algorithm

3.1 Is cosine similarity a good measure by itself?

The cosine measure makes perfect sense for the K-means algorithm. K-means tries to maximize the average pairwise similarity between documents within clusters. The overall pairwise similarity of a cluster is equal to the square of the norm of the centroid vector of the cluster if the cosine measure is used for similarity [4]. Each document is assigned to the cluster whose centroid is most similar to the document, which means that the average similarity between pairs of documents in the cluster is maximized.

In the case of hierarchical clustering, cosine similarity turns out not to be very suitable. For example, for the LA1 document set [17], a document's closest neighbor actually belongs to a different class 20% of the time. In such a scenario, hierarchical methods make many mistakes initially, and these mistakes can never be corrected, at least with standard hierarchical techniques. Thus, while hierarchical clustering techniques are often thought to be the highest quality clustering approach, in general, variants of the K-means algorithms have been found to work as well or better for document clustering [16].

3.2 Our Clustering Algorithm

We begin by calculating the document similarity matrix, i.e., the matrix which gives the cosine similarity for each pair of documents. Once this similarity matrix is calculated, we find the first n nearest neighbors for each document. (Every document is considered to be its own 0th neighbor.) In the nearest neighbor graph, there is a link from document i to document j, if i and j both have each other in their nearest neighbor list. In the shared nearest neighbor graph, there is a link from i to j if there is a link from i to j in the near neighbor graph. The strength of this link is equal to the number of shared near neighbors of i and j.

At this point, we could just apply a threshold, and take all the connected components of the shared nearest neighbor graph as our final clusters [8]. However, this threshold would need to be set too high, since this is a single link approach, and would give poor results when patterns in the data set are not very significant. When a high threshold is applied, a natural cluster might be split into many small clusters due to variations in tightness in the similarity within the cluster. We address these problems with the clustering algorithm described below.

There are two types of parameters used in this algorithm: one type relates to the strength of the links in the shared near neighbor graph, the other type relates to the number of strong links for a document. If the strength of a link is greater than a threshold, that link is labelled as a strong link.

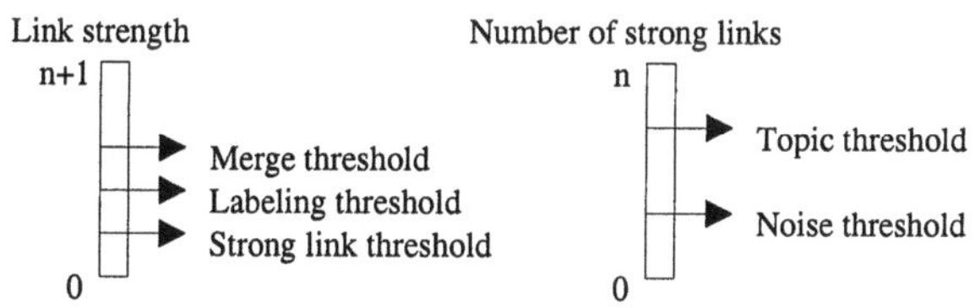

Figure 1: Types of Thresholds for SNN Clustering

The details of our shared nearest neighbor clustering algorithm are as follows:

1. For every point i in the dataset, calculate the connectivity, conn[i], the number of strong links the point has.

2. For a point i in the dataset, if conn[i] $<$ *noise threshold*, then that point is not considered in the clustering since it is similar to only a

few of its neighbors. Similarly, if conn[i] > *topic threshold*, then that point is similar to most of its neighbors and is chosen to represent its neighborhood.

3. For any pair of points (i, j) in the dataset, if i and j share significant numbers of their neighbors, i.e., the strength of the link between i and j is greater than the *merge threshold*, then they will appear together in the final clustering if either one of them (or both) is chosen to be a representative. Note that the algorithm will not suffer from the effects of transitivity since every other point on a chain of links has to be chosen to be a representative. In other words, two documents that are not directly related will be put in the same cluster only if there are many other documents between them that are connected with strong links, half of which must represent their own neighborhood.
4. Labeling step: Having defined the representative points and the points strongly related to them, we can bring back some of the points that did not survive the *merge threshold.* We do this by scanning the shared near neighbor list of all the points that are part of a cluster, and checking whether those points (a) have links to points that don't belong to any cluster and (b) have a link strength greater than the labeling threshold.

The method described above finds communities of documents, where a document in a community shares a certain fraction of its neighbors with at least some number of neighbors. While the probability of a document belonging to a class different from its nearest neighbor's class may be relatively high, this probability decreases as the two documents share more and more neighbors. This is the main idea behind the algorithm.

3.3 Impact of the Size of the Nearest Neighbor List

The size of the near neighbor list, n, is a measure of how focused the clusters will be. It is equal to the smallest number of documents within which we can find a coherent set of documents. If n is equal to 1, then we will end up finding pairs of documents, which are each other's closest neighbors. If n is equal to, say, 50, we could expect to find coherent sets of documents of size 1 to several hundreds. Note that transitivity comes into picture since the requirement for a document to belong to a community is that it must share several neighbors with many other documents in the community, although not necessarily with all of them. Singleton clusters should not be treated as noise, since they were chosen to represent their neighborhood, i.e., there are themes associated with singleton clusters.

Changing the value of n will change the clusters that are found. When n is increased, two points that did not share many neighbors, might now share relatively more neighbors, due to the increased size of the neighbor list. On the other hand, two points that shared a lot of neighbors may now share relatively fewer neighbors.

The nearest neighbor list size should depend on the data set size. For example, if n is fixed and there are infinitely many data points, then the resulting clusters will consist of points that are identical to each other.

3.4 Calculating Link Strengths – a Weighted Approach

Not all the shared nearest neighbors of two points are equally good. If two points share neighbors that are high in their nearest neighbor lists, then these neighbors should make a higher contribution to the similarity between the two points as compared to the case where two points share neighbors on the bottom of their neighbor lists [8]. One disadvantage of the non-weighted scheme is that, when the nearest neighbor list size is increased, all the points start looking the same. In the extreme case, when n is equal to the number of documents less 1, every point has exactly n shared nearest neighbors with every other point. The weighted scheme takes care of this problem to a certain extent and is also more intuitive.

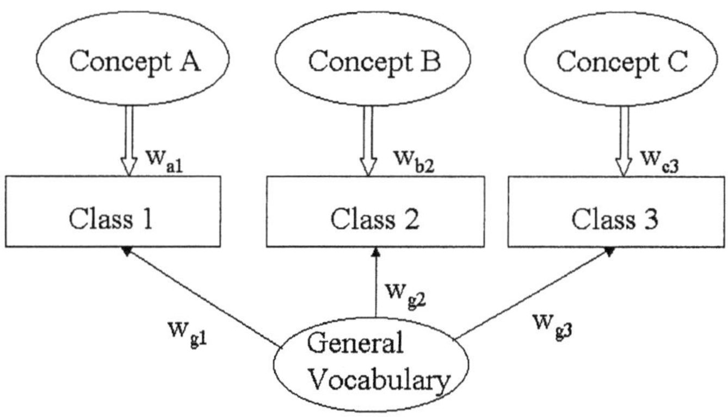

Figure 2: Abstract Data Model

4 Our Abstract Data Model

To enhance our understanding and to better test our algorithm (via synthetic data sets), we developed an abstract data model. In the abstract data model, there are concepts from which the documents "pick" their words. Each concept has a size (the number of words in the concept) and a frequency distribution for the words in the concept. A class has a size (the number of documents belonging to the class), and a length distribution for the documents that belong to the class. Furthermore, a class has links to the concepts from which its documents pick up certain fraction of its words (which is specified by the strength of the link). A word in a concept does not appear in another concept. Overlapping concepts can be modelled by, creating another concept from the intersection of the two concepts, and adjusting the weights accordingly. A simple model that consists of 3 classes and 4 concepts (general vocabulary can be treated as another concept) is shown in Figure 2.

There are several ways to change how "tight" the concepts will be and how "close" they will be to each other. One way is to change the weights of the links from the classes to the concepts. Another way is to change the sizes of the concepts and the general vocabulary.

Consider some scenarios. If the weights of the links from the classes to the general vocabulary are low, then the classes are well separated, since they won't share many words, given that the words in the concepts are distinct. However, even if the weights of the links to the general vocabulary are low, e.g., if the sizes of the concepts are very large compared to the general vocabulary (unlikely scenario), then the classes may not be well separated.

5 Experimental Results

5.1 A Synthetic Data Set Example

In this example there are 9 classes (0-8) and 10 concepts ($a-j$). All concepts except e and j have 30 words, and they appear with a frequency of at most 10 in any document. There are 200 words in concepts e and j, and they appear in a document with a frequency of at most 20. We can think of the concepts $a-e$ as sports vocabularies, concept e being a general sports vocabulary while concepts $a-d$ contain words specific to different sports. Class 0 picks 30% of its words from concept a, 50% of its words from concept e, and its remaining words from the union of all concepts. Similarly, class 1 picks its words from b and e, class 2 picks its words from c and e, and so forth. The same structure exists between classes 4-7 and the concepts $f - j$. Class 8 picks all of its words from the union of the concepts, and is considered to be 'noise.' All the documents contain $20 - 100$ distinct words. Classes 0, 1, 2, and 3 have 100, 200, 300 and 400 documents, respectively. Similarly, classes 4, 5, 6, and 7 have 100, 200, 300 and 400 documents respectively. Class 8 has 1000 documents.

Table 1 below shows how K-means clusters our synthetic data. The first two columns show the cumulative size of the clusters versus the cumulative misclassification, while the next 9 columns are the confusion matrix. The clusters are sorted according to the norm of their centroids, which is given in the last column. (Recall again, that the norm of a cluster centroid represents the overall pairwise document similarity of the cluster.) In this example, 120 clusters were used, but only first 15 are shown.

Table 1: K-means Clustering of Synthetic Data.

cumulative											
size	**mis.**	**0**	**1**	**2**	**3**	**4**	**5**	**6**	**7**	**8**	**norm**
4	2	-	-	-	2	-	-	-	-	2	0.591
12	5	-	5	-	1	-	-	-	-	2	0.488
20	9	-	-	-	-	-	-	-	3	1	0.477
28	9	-	-	-	-	-	-	-	-	8	0.471
36	9	-	-	-	-	-	-	-	-	8	0.466
60	12	-	21	1	-	-	-	-	-	2	0.465
91	14	-	-	-	-	-	29	-	-	2	0.465
110	19	-	-	-	-	2	14	1	-	2	0.463
161	19	-	-	-	-	-	51	-	-	-	0.463
179	23	-	14	-	2	-	-	-	-	2	0.461
226	24	-	-	-	46	-	-	-	-	1	0.459
242	27	-	-	-	13	-	-	-	-	3	0.459
270	29	-	-	-	26	-	-	-	-	2	0.459
295	32	1	-	1	22	-	-	-	-	1	0.458
304	32	-	-	-	-	-	-	-	-	9	0.458
...	...	...	...	...	...	...	...	...	...	...	...
3000	**480**	-	-	-	-	-	-	-	-	15	0.365

Tables 2 and 3 show the clustering results at two different resolutions,

Table 2: SNN Clustering of Synthetic Data - Low Resolution.

0	1	2	3	4	5	6	7	8	norm
83	183	261	351	-	-	-	-	1	0.354
-	-	-	-	89	168	266	339	1	0.354

Table 3: SNN Clustering of Synthetic Data - High Resolution.

0	1	2	3	4	5	6	7	8	
78	-	-	-	-	-	-	-	-	0.459
-	169	-	-	-	-	-	-	-	0.438
1	1	228	-	-	-	-	-	-	0.424
-	1	-	280	-	-	-	-	-	0.423
-	-	-	4	-	-	-	-	-	0.674
-	-	-	-	81	-	-	-	-	0.449
-	-	-	-	-	160	-	-	-	0.443
-	-	-	-	-	-	223	-	-	0.426
-	-	-	-	-	-	-	4	-	0.672
-	-	-	-	-	-	1	279	-	0.420

using the method described in this paper. At a low resolution, Table 2, our technique captures the way that the classes of documents are generated, putting documents from classes 0 – 3 in one cluster and documents from classes 4 – 7 in another. At a higher resolution, shown in Table 3, only the documents generated from single concepts are put together. Note that there are only 2 misclassified documents in the low-resolution case, and only 4 misclassified documents for the high-resolution results. Also, note that documents from the noise cluster are almost completely ignored. We also observed that K-means clusters tend to contain more of the 'general sports vocabulary' (e) in their most important word list, whereas SNN clusters contain the terms specific to their sport (a, b, c, d).

If we know the right number of clusters before running K-means, the results look better, but still not as good as for the SNN clusters. Besides, the right number of clusters is often difficult to determine for K-means.

5.2 Comparison with K-means on a Real Data Set

Data set LA1 is from the Los Angeles Times data of TREC-5 [17]. The words in the tables are the most important 6 words in each document. We see that all the documents in the first cluster are related to NCAA, while all the documents in the second cluster are related to NBA. Even though both sets of documents are basketball related, our clustering algorithm found them as separate clusters. We ran the K-means algorithm on the same data set, and interestingly, all of the documents in these two clusters appeared in the same K-means cluster together with some unrelated documents, including a number of documents related to gymnastics and swimming. The reason that K-means put all these sports documents in the same cluster is that sports documents tend to share a lot of common words, such as *score*, *half*, *quarter*, *game*, *ball*, etc. This example shows that pair-wise similarity, by itself, isn't a good measure for clustering documents.

Table 4: The NCAA Cluster.

wolfpack	towson	lead	tech	Scor	North
syracus	scor	georgia	dome	auburn	Louisvill
Scor	lead	throw	half	Free	Iowa
Scor	Fresno	unlv	lead	lockhart	jacksonvil
Panther	pittsburgh	sooner	brookin	Scor	Game
Iowa	minnesota	scor	illinoi	wisconsin	Burton
Scor	half	virginia	georgetown	lead	Kansa
Burson	louisvill	scor	ohio	game	Ellison

Table 5: The NBA Cluster.

Pacer	scor	piston	shot	game	hawkin
Cavali	mckei	charlott	scor	superson	cleveland
Scor	game	tripucka	basket	hornet	straight
levingston	hawk	jordan	malon	buck	quarter
daugherti	piston	warrior	cavali	shot	Eject

The Tables 6 and 7 show the confusion matrices for the K-means and SNN techniques, respectively, on the LA1 data set, which has the class labels: *Financial*, *Foreign*, *National*, *Metro*, *Sports*, *Entertainment*.

We see, using the SNN approach, that we can get purer clusters, but that

Table 6: Performance of K-means on LA1 Data Set.

cumulative							
size	mis.	**Fin.**	**For.**	**Nat.**	**Met.**	**Sports**	**Ent.**
11	3	1	8	1	1	-	-
94	3	83	-	-	-	-	-
149	3	-	-	-	-	55	-
233	3	-	-	-	-	84	-
283	11	2	42	2	4	-	-
325	18	-	-	-	5	35	2
363	23	1	1	-	1	33	2
389	40	9	7	8	2	-	-
475	68	3	58	6	8	7	4
517	84	2	1	1	3	26	9
626	95	-	1	4	3	98	3
755	96	-	-	-	-	128	1
828	135	11	34	16	11	-	1
...	...	...	...	...	...	...	...
...	...	...	...	...	...	...	...
3204	756	3	2	9	10	1	5

Table 7: Performance of SNN on the LA1 Data Set.

cumulative							
size	mis.	**Fin.**	**For.**	**Nat.**	**Met.**	**Sports.**	**Ent.**
45	0	45	-	-	-	-	-
66	2	19	-	2	-	-	-
74	3	7	-	1	-	-	-
80	3	6	-	-	-	-	-
85	3	5	-	-	-	-	-
90	3	5	-	-	-		
95	5	3	1	-	1	-	-
119	6	-	23	-	1	-	-
137	6	-	18	-	-	-	-
153	6	-	16	-	-	-	-
179	13	2	-	19	1	-	4
190	17	-	1	3	7	-	-
197	18	-	-	1	6	-	-
238	21	-	-	1	2	38	-
267	21	-	-	-	-	29	-
288	21	-	-	-	-	21	-
309	21	-	-	-	-	21	-
326	21	-	-	-	-	17	-
335	21	-	-	-	-	9	-
400	21	-	-	-	-	-	65

not all the documents are assigned to clusters. In order to make a fair comparison, we decided to remove, from K-means clusters, all documents that were relatively dissimilar to the centroid of their cluster. The misclassification rate improved only slightly - it dropped from 23.6% to approximately 20% after throwing away half of the documents.

When we looked at the individual documents in a "supposedly not so good" SNN cluster, we found that the documents do form a coherent group even though they have different class labels. Table 8 below shows an SNN cluster of size 23, where 6 of the documents have a different label than 'metro'. If we look at the top 10 words in each of the documents, we can see that all the documents are about fire hazards. If we look at the second document in the list, we can see that there was a fire in a school in Chile. Since it was in Chile, the article appeared in the *Foreign* section of the newspaper. When we performed the same investigation on the K-means clusters, we found that there are actually several threads of stories in the 'not so good' K-means clusters. This artifact can be attributed to the self-similarity effect as discussed in the following.

Table 8: "Bad" SNN Cluster.

Bad Cluster Sample											
Document Label		**Top 10 words in the documents**									
Metro		tree	christma	malfunct	koenig	fire	sullivan	blaze	firefight	glen	spark
Foreign		chile	fire	eighti	eleven	school	confusion	fled	santiago	upi	eve
National		remer	ren	fire	children	victim	couche	adult	stove	woke	anchorag
Metro		fire	duct	jail	smoke	middleton	rubbish	burn	rise	waft	incid
Metro		fire	alameda	inhal	plate	blaze	coron	oakland	smoke	occup	ident
National		fire	smolder	charlott	cigarett	upi	swept	famili	carolina	injur	di
Metro		palmer	fire	build	melodi	toni	fullerton	in	restaur	furnitur	reopen
National		fire	rome	children	frame	blaze	kill	georgia	upi	swept	burn
Metro		bucher	fire	collag	prado	mccarti	apparatu	casa	watercolor	celebr	depart
Metro		sister	verduzco	fire	ornela	blaze	house	neighbor	angelica	bedroom	di
Metro		apart	fire	granada	serrania	complex	casa	anaheim	karrie	blaze	awaken
Metro		lonczak	lytle	fire	forest	acre	covina	creek	brush	cigarett	blaze
Metro		fire	canyon	madlock	trabuco	wind	emori	firefight	joplin	keener	neal
Metro		alarcon	lojacono	apart	fire	inhal	cate	damag	smoke	joseph	orang
Metro		penelop	griffith	spaniel	blaze	springer	dog	peggi	fire	pet	burn
National		outward	yonker	room	mclaughlin	door	fire	uncertain	blaze	nurs	badli
National		apart	fire	richardson	blaze	kill	upi	tex	heavili	cause	damag
Metro		tang	bird	fire	parrot	feenei	griebe	chilli	heater	anaheim	blaze
Metro		weisenberg	griebe	fire	carport	divorc	car	anaheim	estrang	blood	wife
Metro		fire	arson	duma	grove	garden	gasolin	damag	travel	destroi	busi
Metro		church	williamson	firefight	arson	fire	blaze	incendiari	douse	etern	baptist
Metro		apart	arson	fire	anaheim	complex	blaze	caus	carport	probabl	cross
Metro		womble	fire	neighbor	house	demmer	spence	pendleton	clif	blaze	halloween

Suppose that there is a cluster that contains two totally unrelated documents. If we calculate the similarity of each document to the centroid, we will get a value of 0.5 since the documents themselves constitute half of the centroid, i.e., all the similarity of each document to the centroid is similarity with that part of the centroid that represents the document itself, or self-similarity. If, however, the cluster contains two coherent threads of stories, instead of two single documents, then the situation will still be very similar to the two-document case. Due to the self-similarity effect, K-means is not able to distinguish between a loose cluster and a cluster containing documents from several tight subclusters, representing several threads of stories. This is a problem because the average pairwise similarity, i.e., centroid norm, of coherent clusters varies a lot, typically from 0.25 to 0.60. This situation is exactly what we observed on the K-means clusters from the LA1 data set.

5.3 Word Clusters

Using the same dataset (LA1), we clustered the words instead of the documents. When we cluster the documents and look at the top terms in the centroid of the clusters, we get an idea of what the topic of the cluster is. When we cluster the words, which we do by transposing the document-term matrix and using the exact same algorithm as for clustering documents, we obtain coherent sets of words that form concepts. By contrast, the most important terms in document clusters (topics) may contain several concepts. The concepts found are specific to the dataset used. Here are some concepts that are found by the algorithm in LA1.

```
afghanistan embassi guerrilla kabul moscow rebel soviet troop ussr
withdraw

arab araf israe israel palestinian plo territory

chancellor chemic export german germani kadafi kohl libya libyan
plant poison weapon west

able ago associ believ bit bylin call com consid couldn dai datelin
didn do doesn don experi feel front get graphic gui happen haven
help hope includ isn life little ll look lot major maybe mind month
own people photo probabl re reason recent seen sit soon staf start
success tell time tough tri try ve wasn week wouldn writer

ahead ball basket beate brea chri coach coache consecut el final
finish foul fourth free game grab half halftim hill host jef lead
league led left los lost minut miss outscor overal plai player pointer
quarter ralli rank rebound remain roundup scor score scorer season
```

```
shot steal straight streak team third throw ti tim trail victori
win won

ab bengal bowl cincinnati craig dalla denver esiason field football
francisco giant jerri joe miami minnesota montana nfl oppon pass
pittsburgh quarterback rice rush super table taylor terri touchdown
yard
```

When we look at the word clusters, they form a coherent set. The words in the 1st cluster are all related to the USSR-Afghanistan conflict. The 2nd cluster is about the Arabs, Israelis, and the Palestinians, while the 3rd cluster is about the German-Libyan relationships. The 4th cluster represents the general vocabulary in our abstract data model since it contains generic terms that could appear together in any document. The 5th cluster represents general sports terms, while the 6th cluster contain only the words related to football. As we can see, the word clusters correspond to the concepts in the dataset. Concepts are related to the top words in a document cluster, but there's no one-to-one correspondence. We can expect to have words from several concepts in the list of the top words of a document cluster, since a topic may have words from several concepts. Thus, while our abstract data model is a very simple model, it captures the nature of the text data in some important respects.

6 Conclusions and Future Work

Our research indicates that clustering based on shared nearest neighbors is a better approach than K-means clustering for finding groups of documents with a strong, coherent topic or theme. To explain and understand these results, we introduced a concept-based document model, where each document is generated by choosing its words from a small number of specialized vocabularies plus a general vocabulary, and where the probabilities with which words are chosen from each vocabulary depend on the class (topic) to which the document belongs. This model provides a solid foundation for the work in this paper (and future work) by providing a framework that explains **(a)** how it is possible for two documents to be most similar to each other, but yet be in different classes and **(b)** why a shared nearest neighbor clustering approach might work better than K-means or hierarchical clustering approaches, both of which are based on pairwise similarities.

Our future works relates to two areas: understanding and extending our document models and implementation. In the current implementation of the algorithm, topic and noise thresholds are picked as percentages of the total number of data points and the remaining thresholds are picked as percentages of the number of links in the shared nearest neighbor graph. For example, if we have an idea about the amount of noise in the data, we could set the noise threshold accordingly. Using the same set of parameters, we obtain different link strength thresholds for different datasets since they depend on the structure of the data; they are not specified as pre-set values. While this method of selecting the parameters works a lot better than setting fixed thresholds, it is not fully automatic. Fully automating the selection of parameters requires better understanding of text data.

In terms of further developing our document models, we hope to extend our concept models to obtain a better implementation and to more thoroughly understand the behavior of a wide variety of clustering algorithms on a wide variety of data sets. For example, transaction data, e.g., customer purchasing data, is very similar to document data.

References

[1] Charu C. Aggarwal, Stephen C. Gates and Philip S. Yu, "On the merits of building categorization systems by supervised clustering,"Proceedings of the fifth ACM SIGKDD International Conference on Knowledge Discovery and Data Mining, Pages 352 – 356, 1999.

[2] Douglas R. Cutting, David R. Karger, Jan O. Pedersen, and John W. Tukey, "Scatter/Gather: A Cluster-based Approach to Browsing Large Document Collections", ACM SIGIR '92, Pages 318 – 329, 1992.

[3] Richard C. Dubes and Anil K. Jain, *Algorithms for Clustering Data*, Prentice Hall, 1988.

[4] Inderjit S. Dhillon and Dharmendra S. Modha, "Concept Decompositions for Large Sparse Text Data using Clustering," to appear in Machine Learning, 2000 (also appears as IBM Research Report RJ 10147 (95022), July 8, 1999).

[5] A. El-Hamdouchi and P. Willet, "Comparison of Hierarchic Agglomerative Clustering Methods for Document Retrieval," The Computer Journal, Vol. 32, No. 3, 1989.

[6] K. C. Gowda and G. Krishna, (1978), “Agglomerative Clustering Using the Concept of Mutual Nearest Neighborhood”, Pattern Recognition, Vol. 10, pp. 105-112.

[7] Sudipto Guha, Rajeev Rastogi, and Kyuseok Shim, (1998), “ROCK: A Robust Clustering Algorithm for Categorical Attributes,” In Proceedings of the 15th International Conference on Data Engineering, 1999.

[8] R. A. Jarvis and E. A. Patrick, “Clustering Using a Similarity Measure Based on Shared Nearest Neighbors,” IEEE Transactions on Computers, Vol. C-22, No. 11, November, 1973.

[9] George Karypis and Eui-Hong (Sam) Han , “Concept Indexing: A Fast Dimensionality Reduction Algorithm with Applications to Document Retrieval & Categorization,” CIKM 2000.

[10] George Karypis, Eui-Hong Han, and Vipin Kumar, (1999) “CHAMELEON: A Hierarchical Clustering Algorithm Using Dynamic Modeling,” IEEE Computer, Vol. 32, No. 8, August, 1999. pp. 68-75.

[11] Gerald Kowalski, *Information Retrieval Systems – Theory and Implementation*, Kluwer Academic Publishers, 1997.

[12] L. Kaufman and P. J. Rousseeuw, *Finding Groups in Data: an Introduction to Cluster Analysis*, John Wiley and Sons, 1990.

[13] Daphe Koller and Mehran Sahami, “Hierarchically classifying documents using very few words,” Proceedings of the 14th International Conference on Machine Learning (ML), Nashville, Tennessee, July 1997, Pages 170-178.

[14] Bjorner Larsen and Chinatsu Aone, “Fast and Effective Text Mining Using Linear-time Document Clustering,” KDD-99, San Diego, California, 1999.

[15] C. J. van Rijsbergen, *Information Retrieval*, Buttersworth, London, second edition,1989.

[16] Michael Steinbach, George Karypis, and Vipin Kumar, “A Comparison of Document Clustering Algorithms,” KDD-2000 Text Mining Workshop, 2000.

[17] TREC: Text REtrieval Conference. http://trec.nist.gov

[18] Oren Zamir, Oren Etzioni, Omid Madani, Richard M. Karp, "Fast and Intuitive Clustering of Web Documents," KDD '97, Pages 287-290, 1997.

[19] "Evaluation of Hierarchical Clustering Algorithms for Document Datasets," Ying Zhao and George Karypis, CIKM 2002.

Clustering and Information Retrieval (pp. 105-133)
W. Wu, H. Xiong and S. Shekhar(Eds.)

On Quantitative Evaluation of Clustering Systems

Ji He*, Ah-Hwee Tan‡, Chew-Lim Tan*, Sam-Yuan Sung*

**School of Computing, National University of Singapore*
3 Science Drive 2, Singapore 117543
E-mail: {heji,tancl,ssung}@comp.nus.edu.sg

‡Laboratories for Information Technology
21 Heng Mui Keng Terrace, Singapore 119613
E-mail: ahhwee@lit.org.sg

Contents

1 Introduction

Clustering refers to the task of partitioning unlabelled data into meaningful groups (clusters). It is a useful approach in data mining processes for identifying hidden patterns and revealing underlying knowledge from large data collections. The application areas of clustering, to name a few, include image segmentation, information retrieval, document classification, associate rule mining, web usage tracking, and transaction analysis.

While a large number of clustering methods have been developed, clustering remains a challenging task as a clustering algorithm behaves differently depending on the chosen features of the data set and the parameter values of the algorithm [10]. Therefore, it is important to have some objective measures to evaluate the clustering quality in a quantitative manner. Given a clustering problem with no prior knowledge, a quantitative assessment of the clustering algorithm serves as an important reference for various tasks, such as discovering the distribution of a data set, identifying the clustering paradigm that is most suitable for a problem domain, and deciding the optimal parameters for a specific clustering method.

In the rest of this chapter, we review the process of clustering activities and discuss the factors that affect the output of a clustering system. We then describe two sets of quality measures for the evaluation of clustering algorithms. While the first set of measures evaluate clustering outputs in terms of their inherent data distribution, the second set evaluates output clusters in terms of how well they follow known class distribution of the problem domain. We illustrate the application of these evaluation measures through a series of controlled experiments using three clustering algorithms.

2 Clustering Process: A Brief Review

Jain et al summarized a typical sequence of the clustering activities as the following three stages depicted in Figure 1 [13, 14]:

1. *pattern representation*, optionally including feature extraction and/or selection,
2. definition of a *pattern proximity measure* for the data domain, and
3. *clustering* or *grouping* of data points according to the chosen pattern representation and the proximity measure.

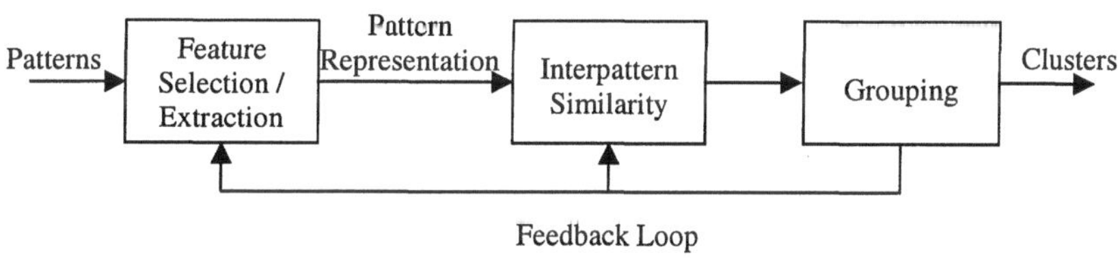

Figure 1: A typical sequencing of clustering activity.

Since the output of a clustering system is the result of the system's interactive activities in each stage, various factors in each stage that affect the system's activity in turn have impact on the clustering output. We extend our discussion in the following subsections.

2.1 Pattern Representation, Feature Selection and Feature Extraction

Pattern representation refers to the paradigm for observation and the abstraction of the learning problem, including the type, the number and the scale of the features, the number of the patterns, and the format of the feature representation. Feature selection is defined as the task of identifying a set of most representative subset of the natural features (or transformations of the natural features) to be used by the machine. Feature extraction, on the other hand, refers to the paradigm for converting the observations of the natural features into a machine understandable format.

Pattern representation is considered as the basis of machine learning. Since human accessibility of the patterns is highly dependent on their representation format, an unsuitable pattern representation may result in a failure of producing meaningful clusters as a user desires. As show in Figure 2a, using a cartesian coordinate representation, a clustering method would have no problem in identify the five compact groups of data points. However, when the same representation is applied to the data set in Figure 2b, the four string-shape clusters would probably not be discovered as they are not easily separable, in terms of Euclidean distance. Instead, a polar coordinate representation could lead to a better result, as the data points in each string-shape cluster are close to each other, in terms of polar angle.

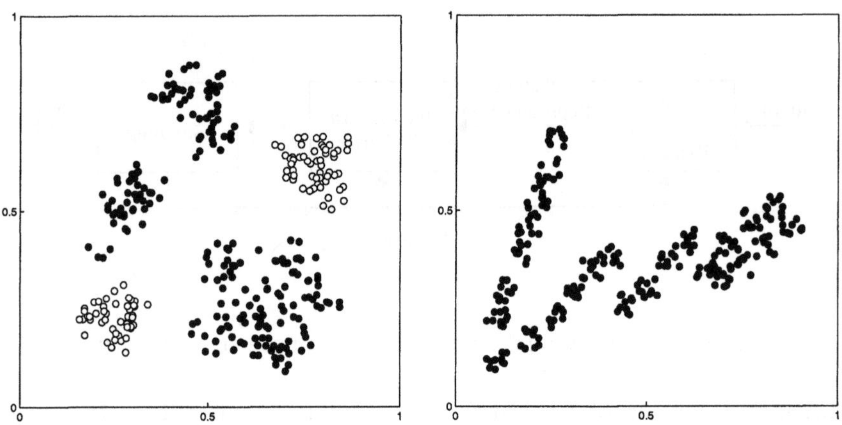

(a) (b)

Figure 2: To identify compact clusters, a cartesian coordinate representation is more suitable for case (a), while a polar coordinate representation is more suitable for case (b).

Feature selection and extraction play an important role for abstracting complex patterns into a machine understandable representation. The feature set used by a clustering system regularizes the "area" that the system gives "attention" to. Referring to the data set in Figure 2a, if *coordinate position* is selected as the feature set, many clustering algorithms would be capable of identifying the five compact clusters. However, if only the *color* of the data points is selected as the feature, a clustering system would probably output only two clusters, containing *white points* and *black points* respectively.

The feature set also affects the quality as well as the efficiency of a clustering system. A large feature set containing numerous irrelevant features does not improve the clustering quality but increases the computational complexity of the system. On the other hand, an insufficient feature set may decrease the accuracy of the representation and therefore cause potential loss of important patterns in the clustering output.

2.2 Pattern Proximity Measure

Pattern proximity refers to the metric that evaluates the similarity (or in contrast, the dissimilarity/distance) between two patterns. While a number of clustering methods (such as [17]) disclaim the use of specific distance measures, they use alternative pattern proximity measures to evaluate the so-called *relationship* between two patterns. A pattern proximity measure serves as the basis for cluster generation as it indicates how two patterns "look alike" to each other.

Since the type, the range, and the format of the input features are defined during the pattern representation stage, it follows that a pattern proximity measure should correspond to the pattern representation. In addition, a good proximity measure should be capable of utilizing only the key features of the data domain. Referring to Figure 2 again, with a cartesian representation, Euclidean distance is suitable to identify the geometric differences among the five clusters in data set (a) but may not be capable enough to recognize the clusters in data set (b). Instead, cosine distance is more suitable for data set (b), as it gives no weight to a vector's radius and focuses on the differences of the vectors' projections on the polar angle only. Generally, a careful review on the existing correlations among patterns helps to choose a suitable pattern similarity measure.

Given an existing pattern representation paradigm, a data set may be separable in various ways. Under this condition, using different pattern proximity measures may result in very different clustering outputs. Figure 3 depicts a simple example using eight speed cameras on three roads. Based on different proximity criteria listed in Table 1, there are different solutions for clustering the speed cameras, each with an acceptable interpretation. In most cases, the clustering system is desired to output only one (or a few number of) optimal grouping solution that best matches the user intention on the data set, although that may be partial and subjective. Hence it is

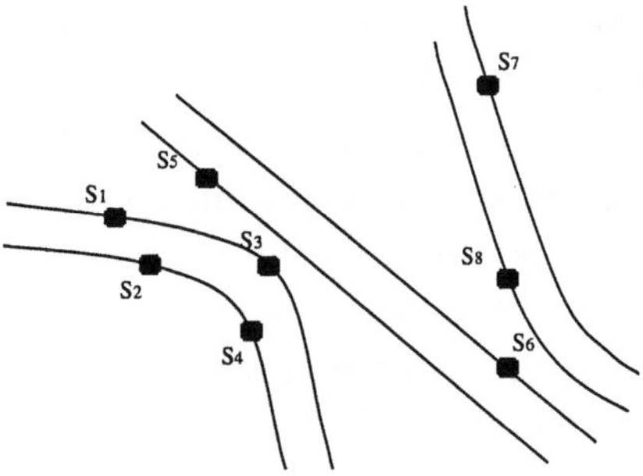

Figure 3: Using various pattern proximity measures, the eight speed cameras on the three roads may be clustered into different cluster groupings.

Table 1: Interpretations to the various clustering results of the eight speed cameras, based on different pattern proximity measures.

Measure	Clustering Result	Interpretation
Geometric distance	$C_1 = \{S_1, S_2, S_3, S_4, S_5\}$ $C_2 = \{S_6, S_8\}$ $C_3 = \{S_7\}$	Cameras in each cluster are geometrically closer to each other than to those in other clusters.
Connectivity	$C_1 = \{S_1, S_2, S_3, S_4\}$ $C_2 = \{S_5, S_6\}$ $C_3 = \{S_7, S_8\}$	Each cluster contains the cameras in the same road.
Density	$C_1 = \{S_1, S_2, S_3, S_4, S_5\}$ $C_2 = \{S_6, S_7, S_8\}$	C_1 identifies the zone intensively equipped with cameras, in contrast to the rest of the area.

important to identify the pattern proximity measure that effectively and precisely formulates the user's intention on the patterns.

2.3 Clustering Algorithms

A clustering algorithm groups the input data according to a set of predefined criteria. The clustering algorithm used by a system can be either statistical or heuristic. In essence, the objective of clustering is to maximize the intra-cluster similarity and minimize the inter-cluster similarity [22]. A large variety of clustering algorithms have been extensively studied in the literature. While a comprehensive survey of clustering algorithms is not the focus of our study, we give a bird's-eye review of various types of the available algorithms in Table 2.

Table 2: Various types of clustering methods, based on learning paradigm, codebook size, cluster assignment, and system architecture respectively.

Criteria	Categories
Learning paradigm	*Off-line:* Iteratively batch learning on the whole input set. *On-line:* Incremental learning that does not remember the specific input history.
Codebook size (number of output clusters)	*Static-sizing:* The codebook size is fixed. *Dynamic-sizing:* The codebook size is adaptive to the distribution of input data.
Cluster assignment	*Hard:* Each input is assigned with one class label. *Fuzzy:* Each input is given a degree of membership with every output cluster.
System architecture	*Partitioning:* The input space is naively separated into disjoint output clusters. *Hierarchical:* The output tree shows the relations among clusters. *Density-based:* The input data are grouped based on density conditions. *Grid-based:* The spacial input space is quantized into finite sub-spaces (grids) before clustering of each sub-space.

Despite the numerous clustering algorithms available in the literature, there is no single method that can cope with all clustering problems. The choice of the clustering algorithm for a specific task affects the clustering result in a fundamental way. In addition, the learning activity of a large number of clustering algorithms is controlled and hence affected by a set of internal parameters. The optimal parameter set is usually decided through empirical experiments on the specific data set.

3 Evaluation of Clustering Quality

With the summarization of the various factors that affect the result of a clustering system, it is desirable for a clustering system to be capable of providing the necessary feedback to each stage of the clustering process, based on the evaluation and the interpretation of the clustering output. Such a feedback helps to gather prior knowledge on the data distribution, define a suitable pattern proximity measure that matches the problem domain, choose a clustering algorithm as well as decide the optimal parameter setting for the specific algorithm, and therefore lead to an optimal clustering output.

Human inspection on the clustering output may be the most intuitive clustering validation method as it compares the clustering result with the user's intention in a natural way. In fact, human inspection has been widely used to validate the clustering outputs of the controlled experiments on two dimensional data sets. However, human inspection lacks the scalability to high dimensional, large, and complicated problem domains. In addition, manual inspection is always not desirable and not feasible in real-life applications. Therefore, quantitative assessment of clustering quality is of great importance for various clustering applications.

When talking about quantitative measure of the clustering quality, readers should be aware that the definition of the *quality* actually is quite subjective. Given an input data set, users may have varying desires on the knowledge that a clustering system could discover, and therefore give different interpretation of the clustering *quality*. Before a quality measure is applied to evaluate the clustering results generated by two different systems, a study on the *quantitative comparability* of the outputs is fundamentally important. We consider two clustering systems quantitative comparable only if:

- *the pattern representation of the two systems are similar enough,*
- *they are based on the similar clustering fundament,* and
- *they intend to fulfill the same user requirement, optionally to optimize the same criteria.*

We give explanations through examples below. Given the speed camera example as in Figure 3, the three clustering results in Table 1 are not quantitatively comparable as they are based on different clustering criteria. Likewise, given the data set as in Figure 2a, the clustering output based on the *coordinate position* of each data point is not quantitatively comparable with that based on the *color* of each data point, as they are based on totally different feature sets. However, using the same feature representation (*coordinate position*), the outputs of the two clustering methods as in Figure 4 are quantitatively comparable as they both attempt to identify compact clusters of the data set.

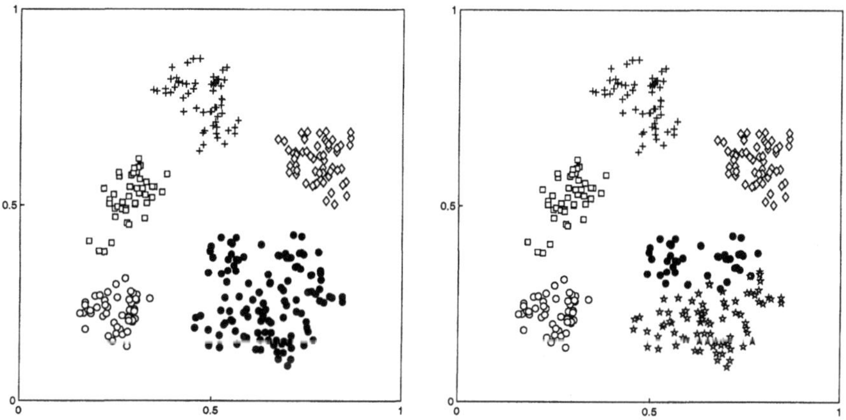

(a) (b)

Figure 4: Two quantitatively comparable partitioning outputs of the data set in Figure 2a. Each type of marker identifies data points in the same cluster. Result (a) is considered to have a higher quality than result (b) in the sense that it recognizes the large cluster (marked with solid circles in (a)) more precisely.

A multitude of clustering evaluation measures have been extensively studied in the literature. Examples include the Dunn and Dunn-like family of measures initially proposed for evaluation of *crisp* [7], the variances of DB measures [6, 16], and the relatively recent SD validity indices [11]. However, due to the limitations mentioned above, a large number of these evaluation measures are capable of validating a narrow ranges of in-house clustering methods only. Our study intents to use a set of evaluation measures with high scalability, in terms of the capability for evaluating a wide range of clustering systems. In the rest of this section, we introduce several clustering evaluation measures based on two different statistical fundaments, i.e. cluster distribution and class conformation respectively.

3.1 Evaluation Measures Based on Cluster Distribution

The objective of clustering has been widely quoted as to reorganize the input data set in an unsupervised way such that data points in the same cluster are more similar to each other than to points in different clusters. When explaining this objective in the quantitative manner, it is to minimize the distances among the data points in individual clusters and to maximize the distances between clusters. Therefore, it is a natural way to validate the intra-cluster homogeneity and the inter-cluster separation of the clustering output in a global fashion, using the quantities inherent to the distribution of the output data.

We extend our study from the various clustering validity methods in this category [9, 11], and propose two quality evaluation measures, namely: *cluster compactness* and *cluster separation*. The definitions of these measures are given as below.

Cluster Compactness: The *cluster compactness* measure introduced in our study, is based on our generalized definition of the *variance* of a vector data set given by

$$v(\mathbf{X}) = \sqrt{\frac{1}{N}\sum_{i=1}^{N} d^2(\mathbf{x}_i, \overline{\mathbf{x}})} \tag{1}$$

where $d(\mathbf{x}_i, \mathbf{x}_j)$ is a distance metric between two vectors $\mathbf{x}_i$ and $\mathbf{x}_j$, N is the number of members in $\mathbf{X}$, and $\overline{\mathbf{x}} = \frac{1}{N}\sum_i \mathbf{x}_i$ is the mean of $\mathbf{X}$. A smaller *variance* value of a data set indicates a higher homogeneity of the vectors in the data set, in terms of the distance measure $d()$. Particularly, when $\mathbf{X}$ is one-dimensional and $d()$ is the Euclidean distance, $v(\mathbf{X})$ becomes the statistical variance of the data set $\sigma(\mathbf{X})$. The *cluster compactness* for the output clusters $\mathbf{c}_1, \mathbf{c}_2, \cdots, \mathbf{c}_C$ generated by a system is then defined as

$$Cmp = \frac{1}{C}\sum_{i}^{C} \frac{v(\mathbf{c}_i)}{v(\mathbf{X})}, \tag{2}$$

where C is the number of clusters generated on the data set $\mathbf{X}$, $v(\mathbf{c}_i)$ is the variance of the cluster $\mathbf{c}_i$, and $v(\mathbf{X})$ is the variance of the data set $\mathbf{X}$.

The *cluster compactness* measure evaluates how well the subsets (output clusters) of the input is redistributed by the clustering system, compared with the whole input set, in terms of the data homogeneity reflected by the distance metric used by the clustering system. When the distance metric is the Euclidean distance, the *cluster compactness* measure becomes coherent to the *average cluster scattering* index used in Halkidi et al's study [11]. It is understandable that, for the *cluster compactness* measure, a smaller value indicates a higher average compactness in the output clusters. This however does not necessarily mean a "better" clustering output. Given a clustering system that encodes each and every unique input data into one separate cluster, the *cluster compactness* score of its output has a minimal value of 0. Such a clustering output is however not desirable. To tackle this, we introduce the *cluster separation* measure to complement the evaluation.

Cluster Separation: The *cluster separation* measure introduced here borrows the idea in [11] and combines the idea of the clustering evaluation function introduced by [9]. The *cluster separation* of a clustering system's output is defined by

$$Sep = \frac{1}{C(C-1)} \sum_{i=1}^{C} \sum_{j=1, j \neq i}^{C} \exp\left(-\frac{d^2(\mathbf{x}_{c_i}, \mathbf{x}_{c_j})}{2\sigma^2}\right), \tag{3}$$

where σ is a Gaussian constant, C is the number of clusters, $\mathbf{x}_{c_i}$ is the centroid of the cluster $\mathbf{c}_i$, $d()$ is the distance metric used by the clustering system, and $d(\mathbf{x}_{c_i}, \mathbf{x}_{c_j})$ is the distance between the centroid of $\mathbf{c}_i$ and the centroid of $\mathbf{c}_j$.

It is noted that the pair-wise distances among the output cluster centroids are the key components of the *cluster separation* measure. The Gaussian function and the L1-normalization normalizes its value to between 0 and 1. A smaller *cluster separation* score indicates a larger overall dissimilarity among the output clusters. However, given the particular case that a clustering system output the whole input set into one cluster, the *cluster separation* score reaches to minimal value of 0, which is not applicably desirable. Hence, we reach the necessary point to combine the *cluster compactness* and *cluster separation* measures into one in order to tackle each one's deficiency and evaluate the overall performance of a clustering system. An intuitive combination, named *overall cluster quality*, is defined as:

$$Ocq(\beta) = \beta \cdot Cmp + (1 - \beta) \cdot Sep, \tag{4}$$

where $\beta \in [0, 1]$ is the weight that balances measures *cluster compactness* and *cluster separation*. Particularly, $Ocq(0.5)$ gives equal weights to the two measures. Readers however should be aware of the limitation of this combination: although the combination measure Ocq facilitates the comparison work, one may find it not easy to interpret the combined value, as the Cmp and Sep scores are not measured in the same dimension. In addition, in some other cases, evaluating a clustering system through intra-cluster compactness and inter-cluster separation respectively helps to gain more insightful understanding of the system characteristics.

3.2 Evaluation Measures Based on Class Conformation

This category of validation measures assumes that there is a desirable distribution of the data set with which it is possible to perform a direct comparison of the clustering output. Following the data distribution, one can assign a class label to each data point. The target of the clustering system can then be correspondingly interpreted as to replicate the underlying class structure through unsupervised learning. In an optimal clustering output, data points with the same class labels are clustered into the same cluster and data points with different class labels appear in different clusters. We describe two quantitative evaluation measures based on these criteria as follows.

Cluster Entropy: Boley [1] introduced an information entropy approach to evaluate the quality of a set of clusters according to the original class labels of the data points . For each cluster $\mathbf{c}_i$, a cluster entropy Ec_i is computed by

$$Ec_i = -\sum_j \frac{n(l_j, c_i)}{n(c_i)} log \frac{n(l_j, c_i)}{n(c_i)} \tag{5}$$

where $n(l_j, c_i)$ is the number of the samples in cluster $\mathbf{c}_i$ with a predefined label l_j and $n(c_i) = \sum_j n(l_j, c_i)$ is the number of samples in cluster $\mathbf{c}_i$. The overall *cluster entropy* Ec is then given by a weighted sum of individual cluster entropies by

$$Ec = \frac{1}{\sum_i n(c_i)} \sum_i n(c_i) Ec_i. \tag{6}$$

The cluster entropy reflects the quality of individual clusters in terms of homogeneity of the data points in a cluster (a smaller value indicates a higher homogeneity). It however does not measure the compactness of a clustering solution in terms of the number of clusters generated. A clustering system that generates many clusters would tend to have very low cluster entropies but is not necessarily desirable. To counter this deficiency, we use another entropy measure below to measure how data points of the same class are represented by the various clusters created.

Class Entropy: For each class $\mathbf{l}_j$, a class entropy El_j is computed by

$$El_j = -\sum_i \frac{n(l_j, c_i)}{n(l_j)} log \frac{n(l_j, c_i)}{n(l_j)} \tag{7}$$

where $n(l_j, c_i)$ is the number of samples in cluster $\mathbf{c}_i$ with a predefined label l_j and $n(l_j) = \sum_i n(l_j, c_i)$ is the number of the samples with class label l_j. The overall *class entropy* El is then given by a weighted sum of individual class entropies by

$$El = \frac{1}{\sum_j n(l_j)} \sum_j n(l_j) El_j. \tag{8}$$

Since both the *cluster entropy* and *class entropy* utilize the predefined class labels on the input data only, they are independent to the choice of the feature representation and the pattern proximity measure. Therefore these two measures are practically capable of evaluating any clustering system. Compared with the cluster distribution based measures, these class conformation based measures are likely to have more advantages for identifying the optical clustering solution to match the user's intention on the problem domain. One apparent drawback however is the potential complexity of the labelling process, which may not be feasible and not desirable in real-life applications.

Our prior study showed that, similar to the characteristics of the *cluster compactness* and the *cluster separation* measures, as the number of clusters over one data set increases, the *cluster entropy* generally decreases while the *class entropy* increases. We follow the identical paradigm for the combination of *cluster compactness* and *cluster separation*, and define a combined *overall entropy* measure:

$$Ecl(\beta) = \beta \cdot Ec + (1 - \beta) \cdot El, \tag{9}$$

where $\beta \in [0, 1]$ is the weight that balances the two measures.

4 Clustering Methods

Before introducing our experiments on the various clustering quality measures, review of the clustering systems used in our experiments helps toward a better understanding of the experimental results. Our experiments tested three partitioning methods that work with a fixed number of clusters, namely k-means [18], Self-Organizing Maps (SOM) [15], and Adaptive Resonance

Theory under Constraints (ART-C) [12]. The learning algorithms of the three methods are summarized as follows.

4.1 k-means

k-means [18] has been extensively studied and applied in the clustering literature due to its simplicity and robustness. The fundament of the k-means clustering method is to minimize the intra-cluster compactness of the output, in terms of the summed squared error. The k-means clustering paradigm is summarized below.

1. Initialize the k reference clusters with randomly chosen input points or through certain estimation of the data distribution.
2. Assign each input point to the nearest cluster centroid.
3. Recalculate the centroid of each cluster using the mean of the input points in the cluster.
4. Repeat from step 2 until convergence.

Since k-means follows a batch learning paradigm to iteratively adjust the cluster centroids, it is inefficient in handling large scale data sets. In addition, the output of k-means is affected by the cluster initialization method. Its strength however lies in the satisfactory quality in the sense that the output has locally minimal summed squared error when it converges.

4.2 Self-Organizing Maps (SOM)

SOM as proposed by Kohonen [15] is a family of self-organizing neural networks widely used for clustering and visualization. As an unique feature of SOM, the clusters in a SOM network are organized in a multi-dimensional map. During learning, the network updates not only the winner's weight, but also the weights of the winner's neighbors. This results in an output map with a distribution that similar patterns (clusters) are placed together. The learning paradigm of SOM is given below.

1. Initialize each reference cluster $\mathbf{w}_j^{(0)}$ with random values. Set the initial neighborhood set of each cluster $N_j^{(0)}$ to be large.
2. Given an input $\mathbf{x}_i$, find the winner node J that has the maximal similarity with $\mathbf{x}_i$.

3. Update the cluster vectors of the winner node and its neighbors, according to

$$\mathbf{w}_j^{(t+1)} = \mathbf{w}_j^{(t)} + \eta^{(t)} h(j, J)(\mathbf{x}_i - \mathbf{w}_j^{(t)}) \text{ for each } j \in N_J^{(t)}, \tag{10}$$

where $h(j, J) \in [0, 1]$ is a scalar *kernel* function that gives a higher weight to a closer neighbor of the winner node J and $\eta^{(t)}$ is the learning rate.

4. At the end of the learning iteration, shrink the neighborhood sets so that $N_j^{(t+1)} \subset N_j^{(t)}$ for each j and decrease the learning rate so that $\eta^{(t+1)} < \eta^{(t)}$. Repeat from step 2 until convergence.

There are a large number of SOM variances depending on the dimension of the organization map, the definition of the neighborhood sets N_i, the kernel function $h()$, as well as the paradigm that iteratively readjusts N_i and η. When the network utilizes one-dimensional map and the degree of neighborhood is zero (i.e. the neighborhood contains the winner node only), SOM equals to an online learning variance of the k-means clustering method [8]. It is also understandable that, if the neighborhood degree shrinks to zero at the end of the learning, SOM is capable of obtaining optimal output with locally minimal summed square error. The drawback of SOM is that its learning activity is affected by the initialization of network and the presentation order of the input data.

4.3 Adaptive Resonance Theory under Constraints (ART-C)

ART-C [12] is a new variance of the Adaptive Resonance Theory (ART) neural networks, which was originally developed by Carpenter and Grossberg [5]. Unlike the conventional ART modules which work on a dynamic number of output clusters, an ART-C module is capable of satisfying a user *constraint* on its codebook size (i.e. the number of output clusters), while keeping the stability-plastically of ART intact. Compared with the conventional ART architecture, the ART-C architecture (Figure 5) contains an added constraining subsystem, which interacts with the ART's attentional subsystem and the orienting subsystem. During learning, the constraining subsystem adaptively estimates the distribution of the input data and self-adjusts the vigilance parameter for the orienting subsystem, which in turn governs the learning activities in the attentional subsystem. The learning paradigm of ART-C is summarized below.

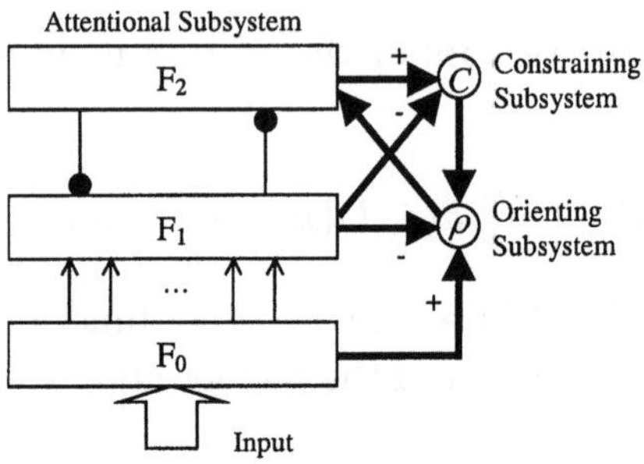

Figure 5: The ART-C architecture.

1. Initialize the network's recognition layer F_2 with null set $\emptyset$ (i.e. the number of clusters in F_2 is zero) and set the network's *vigilance* ρ as 1.0.

2. Given an input $\mathbf{x}_i$ in the input layer F_0, the comparison layer F_1 stores the *match* scores $M(\mathbf{x}_i, \mathbf{w}_j)$ between the input and every cluster vector $\mathbf{w}_j$ in F_2.

3. If the maximal *match* scores satisfies $\max\{M(\mathbf{x}_i, \mathbf{w}_j)\} \geq \rho$ or the number of clusters c in F_2 satisfies $c < C$, where C is the user constraint on the number of output clusters, then carry out the *conventional ART learning* process, which is summarized below,

   (a) Calculate the *choice* scores $T(\mathbf{x}_i, \mathbf{w}_j)$ between the input and every cluster vector $\mathbf{w}_j$ in F_2.

   (b) Identify the winner node J that receives the maximal *choice* score $T(\mathbf{x}_i, \mathbf{w}_J)$. *Resonance* happens when the *match* score $M(\mathbf{x}_i, \mathbf{w}_J) \geq \rho$. Otherwise *reset* the winner node J and repeat the search process.

   (c) If the network reaches *resonance*, the network updates the cluster vector of the winner node, according to a learning function

$$\mathbf{w}_J^{(t+1)} = L(\mathbf{x}_i, \mathbf{w}_J^{(t)}). \tag{11}$$

   Otherwise, if all F_2 nodes j are reset, insert the input $\mathbf{x}_i$ into the F_2 layer as a new reference cluster.

otherwise, do *constraint reset*, which is summarized below,

(a) Insert input $\mathbf{x}_i$ into F_2 layer as a new reference cluster.

(b) Calculate the pair wise *match* score $M(\mathbf{w}_i, \mathbf{w}_j)$ of every F_2 node pairs.

(c) Locate the winner pair (I, J) according to

$$M(\mathbf{w}_I, \mathbf{w}_J) = \max\{M(\mathbf{w}_i, \mathbf{w}_j) : i \in \mathbf{R}\}, \tag{12}$$

where $\mathbf{R}$ is the set of F_2 nodes indexed by the criteria:

$$\mathbf{R} = \{F_2 \text{ node } i \text{ whose } \max\{M(\mathbf{w}_i, \mathbf{w}_j) : j \neq i\} < \rho\}. \tag{13}$$

(d) Modify the network's *vigilance* according to the *match* score between the winner pair:

$$\rho^{new} = M(\mathbf{w}_I, \mathbf{w}_J). \tag{14}$$

(e) Update $\mathbf{w_J}$ with $\mathbf{w_I}$, by utilizing the ART *learning* function:

$$\mathbf{w}_J^{(t+1)} = L(\mathbf{w}_I^{(t)}, \mathbf{w}_J^{(t)}). \tag{15}$$

(f) Delete node I from F_2 layer.

4. Repeat from step 2 until convergence.

The ART module used in the ART-C architecture can be either ART-1 [5], ART-2 [2, 4] or fuzzy ART [3], each using a different set of *choice*, *match*, and *learning* functions. ART-2 utilizes the cosine similarity as the *choice* and *match* functions:

$$T(\mathbf{x}_i, \mathbf{w}_j) = M(\mathbf{x}_i, \mathbf{w}_j) = \frac{\mathbf{x}_i \cdot \mathbf{w}_j}{||\mathbf{x}_i|| ||\mathbf{w}_j||}, \tag{16}$$

where the L_2-norm function $|| \cdot ||$ is defined by

$$||\mathbf{x}|| = \sqrt{\sum_i x_i^2} \tag{17}$$

for vector $\mathbf{x}$. The learning function is given by

$$L(\mathbf{x}_i, \mathbf{w}_j) = \eta \mathbf{x}_i + (1 - \eta) \mathbf{w}_j. \tag{18}$$

where η is the learning rate. As a comparison, fuzzy ART uses hyper-rectangle based *choice* and *match* functions

$$T(\mathbf{x}_i, \mathbf{w}_j) = \frac{|\mathbf{x}_i \wedge \mathbf{w}_j|}{\alpha + |\mathbf{w}_j|}, \tag{19}$$

$$M(\mathbf{x}_i, \mathbf{w}_j) = \frac{|\mathbf{x}_i \wedge \mathbf{w}_J|}{|\mathbf{x}_i|}, \tag{20}$$

where α is a constant, the fuzzy AND operation $\wedge$ is defined by

$$(\mathbf{p} \wedge \mathbf{q})_i \equiv min(p_i, q_i) \tag{21}$$

and the L1-norm $|.|$ is defined by

$$|\mathbf{p}| \equiv \sum_i p_i \tag{22}$$

for vectors $\mathbf{p}$ and $\mathbf{q}$. The learning function for fuzzy ART is given by

$$L(\mathbf{x}_i, \mathbf{w}_j) = \eta(\mathbf{x}_i \wedge \mathbf{w}_j) + (1 - \eta)\mathbf{w}_j. \tag{23}$$

ART-C adaptively generates reference clusters (recognition categories) using the input samples. Therefore no prior knowledge on the data distribution is required for the initialization of the network. However, like most online learning algorithms, ART-C's learning is affected by the presentation order of the input data.

5 Experiments and Discussions

The experiments reported in this section illustrate the use of the various clustering evaluation measures for a variety of purposes. We extend our discussion on the two synthetic data sets as in Figure 2a and 2b in a quantitative manner. Tasks on these two data sets include identifying the optimal number of clusters in the data set and choosing of a pattern proximity measure suitable for the specific data distribution. In addition, using a high-dimensional and sparse real-life data set, i.e. the Reuters-21578 free text collection, we carry out comparisons across the performance of three distinct clustering algorithms, namely k-means, SOM, and ART-C, to discover the similarities and differences of their clustering behaviors.

5.1 Statistical Validation of Comparative Observation

It is noted that, for both k-means and SOM, the clustering output is affected by the initialization of cluster prototypes. In addition, the order of the input sequence affect the outputs of SOM and ART-C. Therefore, comparative findings based on a single experiment would not be representative due to the potential deviation of the observation values.

To tackle this deficiency, we adopted the commonly used statistical validation paradigm in our experiments. Given a clustering task, we repeat the experiments for each clustering method under evaluation for ten times. In each experiment, the presenting sequence of the input data is reshuffled and the clustering methods are trained to convergence. Based on the observation values from the ten runs, the *means* and the *standard deviations* are reported. In order to compare the evaluation scores obtained, we compare the *mean* values and employed t-test to validate the significance of the comparative observations across the ten runs.

5.2 Identification of the Optimal Number of Clusters

Our experiments on the synthetic data set as shown in Figure 2a evaluated the k-means clustering method using the Euclidean distance. The synthetic data set contains 334 data points, each is a 2-dimensional vector of values between 0 and 1. Our task is to identify the optimal number of clusters on the data set in an unsupervised way. Here the *optimal* solution refers to the result that best reflects the data distribution and matches the user's post-validation on the data set, in terms of intra-cluster compactness and inter-cluster separation of the output clusters.

The paradigm of the experiment is summarized below. We apply the k-means method on the data set using k values ranging from 2 to 8. For each k value, we evaluate the quality of the output using the measures based on *cluster compactness* (Cmp) and *cluster separation* (Sep). This enables us to observe the change of the score values according to the change of k. Intuitively the most satisfactory quality score indicates the best partition of the data set, while the corresponding k value suggests the optimal number of clusters on the data set.

Figure 6 depicts the change of *cluster compactness*, *cluster separation*, as well as *overall cluster quality* by varying k from 2 to 8. $2\sigma^2 = 0.25$ is used for the ease of evaluation in Equation 3 and $\beta = 0.5$ for $Ocq()$ is used to give equal weights to *cluster compactness* and *cluster separation*. To obtain a clear illustration, only the mean values of the ten observations over each measure are plotted in the figure while the standard deviations are not reported.

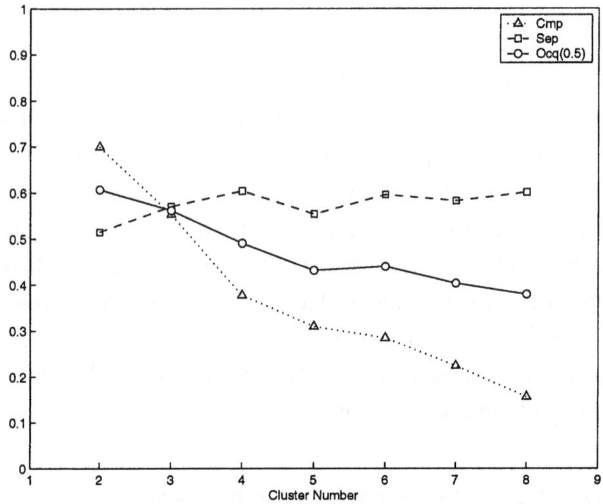

Figure 6: *Cluster compactness*, *cluster separation*, and *overall cluster quality* of k-means on the synthetic data set in Figure 2a. The locally minimal value of *overall cluster quality* $Ocq(0.5)$ at $k = 5$ suggests the optimal number of clusters on the data set.

It is noted that, when k increases, *cluster compactness* gradiently decreases and *cluster separation* generally increases. This is due to the nature that a larger number of partitions on the same data space generally tends to decrease the size of each partition (which causes higher compactness in each partition) as well as the distances among the partition centroids (which causes lower separation of partitions). However, as an apparent exception, the *cluster separation* shows a locally minimal value at $k = 5$, while the decreasing trend of *cluster separation* at $k = 5$ is significantly different from those at different k values. The *overall cluster quality* also shows a locally minimal value at $k = 5$. This suggests that the optimal number of clusters (in terms of Euclidean similarity) is five. The result is supported by human inspection of the data in Figure 2a.

The drawbacks of this experimental paradigm however are notable. First, it is not easy to suggest a proper range of k values for the iterative testing if the user lacks a prior estimation of the data distribution. In addition, both the σ value for the calculation of *cluster separation* and the weight β for the calculation of *overall cluster quality* are subjectively determined. This shows that human interaction and prior knowledge on the problem domain is still required for the use of these evaluation measures.

5.3 Selection of Pattern Proximity Measure

Our experiments on the synthetic data set as shown in Figure 2b utilize the quality measures based on class conformation to evaluate two variances of ART-C networks, namely fuzzy ART-C (based on Fuzzy ART) and ART2-C (based on ART-2). While fuzzy ART-C groups input data according to *nearest hyper-rectangles*, ART2-C groups input data according to *nearest neighbors*, in terms of cosine similarity. Our comparative experiments attempt to discover which of these two pattern proximity measures is more capable of identifying the data distribution on the problem domain, and therefore produces clustering output with a higher match with the user intent.

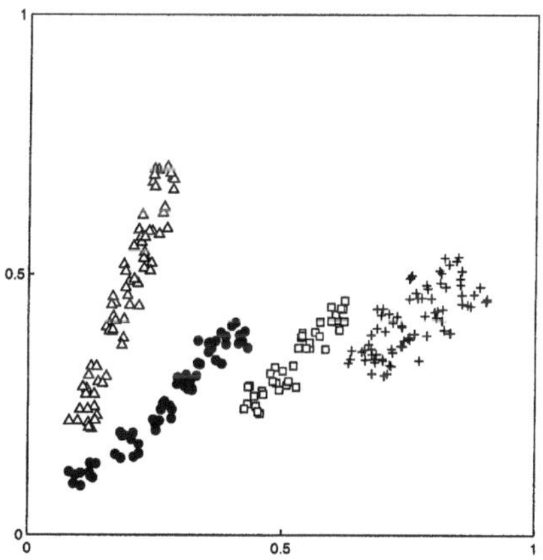

Figure 7: The manually labelled data set as in Figure 2b. Data points assigned with the same class label are identified with the same marker.

In order to evaluate the clustering quality using *cluster entropy*, *class entropy*, and *overall entropy*, we pre-assigned each data point with a class label based on our observation. There are four different class labels assigned to the 250 data points in the collection, each label corresponding to a string-shaped class in Figure 7.

Our experiments compared fuzzy ART-C and ART2-C with a preset constraint (C) of 4, each using a standard set of parameter values. Table 3 summarizes the statistics of the comparison results. While ART2-C is capable of producing a better balanced set of *cluster entropy* and *class entropy* scores (which indicates a better balance of cluster homogeneity and class compactness), the *cluster entropy* score of fuzzy ART-C is three times higher than that of ART2-C in our experiment. Although the *class entropy* score of fuzzy ART-C is slightly lower than that of ART2-C, the weighted *overall entropy* $Ecl(0.5)$ of fuzzy ART-C is significantly higher than that of ART2-C due to the high *cluster entropy* value. This indicates that the cosine similarity based paradigm is more suitable than the nearest hyper-rectangle based paradigm on the tested problem domain. This result is not surprising to us, as prior comparison studies on hyper-rectangle methods also showed that they perform well only when the data boundaries are roughly parallel to the coordinates axes [19].

Table 3: *Cluster entropy*, *class entropy*, and *overall entropy* of ART2-C and fuzzy ART-C on the synthetic data set in Figure 7. Both methods work on $C = 4$. All values are shown with the means and the standard deviations over ten runs.

Method	Ec	El	$Ecl(0.5)$
ART2-C	0.1483 ± 0.0173	0.1142 ± 0.0225	0.1312 ± 0.0092
fuzzy ART-C	0.5800 ± 0.0037	0.0337 ± 0.0064	0.3068 ± 0.0020

5.4 Cross-Comparison of Clustering Methods

We applied the four evaluation measures introduced in this chapter, namely *cluster compactness* (Cmp), *cluster separation* (Sep), *cluster entropy* (Ec), and *class entropy* (El) to compare the performance of k-means, Self-Organizing Maps (SOM), and Adaptive Resonance Theory under Constraints (ART-C) on a sparse and high-dimensional real-life data set, namely the Reuters-21578 free text collection. The details of our benchmark study are reported in the following subsections.

5.4.1 Data Preparation

The Reuters-21578 data set is a collection of documents that appeared on the Reuters news-wire in 1987. Since the data set was originally released for the evaluation of text classification systems, the documents have been carefully assembled and indexed with class labels by personnel from Reuters Ltd. Our experiments used the subset of documents from the top ten categories. To facilitate our evaluation, documents that were originally indexed with multiple class labels were duplicated in our experiment so that each copy was associated with one class label.

We adopted the bag-of-words feature representation scheme for the documents. CHI (χ) statistics [21] was employed as the ranking metric for feature selection. Based on a bag of 335 top-ranking keyword features, the content of each document was represented as an in-document term frequency (TF) vector, which was then processed using an inverse document frequency (IDF) based weighting method [20] and subsequently L2-normalized. After the removal of 57 null vectors (i.e. vectors with all attributes equal to 0), we obtained a set of 9,968 labelled vectors for our experimental study.

5.4.2 Evaluation Paradigm

All three methods used the cosine similarity measure and a standard set of parameter values. As the data collection is large and sparse, we did not expect the clustering methods to replicate the ten clusters corresponding to the ten original classes. Instead, we carried out two set of experiments, each setting the number of the output clusters to be 25 and 81 respectively. In the SOM architecture, these correspond to a 5 by 5, and a 9 by 9 two-dimensional maps.

Unlike the previous two experiments, our experiment on the Reuters-21578 data set evaluated the three clustering methods' performance using each measure separately for a better understanding of their properties. $2\sigma^2 = 1.0$ was used for the ease of computing *cluster separation.* Since the document vectors are preprocessed with L2-normalization, we use Euclidean distance for the evaluation of *cluster compactness* and *cluster separation.* This is due to the high correlation between the cosine similarity and the Euclidean distance under this condition, i.e. high cosine similarity corresponds to close Euclidean distance.

In addition to the four cluster quality measures, the time complexity of each tested system, in terms of the CPU time used on each experiment, were reported and compared. Based on these, we empirically examined the learning efficiency of each method. To facilitate the comparison, all the three systems were implemented with C++ programs that shared a common set of functions for vector manipulation.

5.4.3 Results and Discussions

Table 4 reports the experimental results on k-means, SOM, and ART2-C. Working with 81 clusters, the output of SOM showed a slightly worse quality than that of k-means, in terms of *cluster compactness*, *cluster separation*, and *class entropy*. It may be that the nature of the SOM's learning in maintaining the neighborhood relationship decreases the dissimilarity among clusters as well as the compactness of each cluster. However, the differences were not significantly reflected when the number of output clusters was 25. In general, the evaluation scores of k-means and SOM were rather similar, compared with those of ART2-C. This may be due to the same paradigm we use to initialize the cluster prototypes for both k-means and SOM. More importantly, the learning paradigms of k-means and SOM are more similar to each other than to that of ART2-C. We thus focus our further discussions on the comparison of ART2-C with k-means and SOM.

Table 4: Experimental results for k-means, SOM, and ART2-C on the Reuters-21578 corpus, when the number of clusters were set to 25 and 81 respectively. I and T stand for the number of learning iterations and the cost of training time (in seconds) respectively. Cmp, Sep, Ec and El stand for cluster compactness, cluster separation, cluster entropy and class entropy respectively. All values are shown with the mean and the standard deviation over ten runs.

	Cluster number = 25		
	k-means	SOM	ART2-C
I	10.9 ± 2.0	11.2 ± 1.8	2.2 ± 0.4
T (s)	103.36 ± 36.543	90.018 ± 13.738	42.724 ± 15.537
Cmp	0.4681 ± 0.0086	0.4560 ± 0.0106	0.5187 ± 0.0044
Sep	0.2312 ± 0.0286	0.2248 ± 0.0617	0.2063 ± 0.0143
Ec	0.2028 ± 0.0057	0.2154 ± 0.0122	0.2670 ± 0.0165
El	0.7795 ± 0.0056	0.7838 ± 0.0164	0.7586 ± 0.0341
	Cluster number = 81		
	k-means	SOM	ART2-C
I	11.8 ± 1.6	12.3 ± 1.3	2.8 ± 1.0
T (s)	323.85 ± 74.285	310.84 ± 33.656	88.057 ± 45.738
Cmp	0.3957 ± 0.0116	0.4126 ± 0.0041	0.4594 ± 0.0037
Sep	0.1968 ± 0.0187	0.2135 ± 0.0217	0.1874 ± 0.0243
Ec	0.1808 ± 0.0019	0.1806 ± 0.0030	0.1983 ± 0.0060
El	1.2375 ± 0.0122	1.2592 ± 0.0065	1.1580 ± 0.0127

In our experiments, the *cluster entropy* scores (Ec) and the *cluster compactness* scores (Cmp) of ART2-C outputs were generally higher (which indicated worse data homogeneity within clusters) than those of SOM and k-means. Our explanations are as follows: The learning paradigms of SOM and k-means minimize the mean square error of the data points within the individual clusters. Therefore both SOM and k-means are more capable of generating compact clusters. In terms of *class entropy* (El), and *cluster separation* (Sep), the outputs of ART2-C were better than those of SOM and k-means. It may be that, whereas SOM and k-means modify existing, randomly initialized cluster prototypes to encode new samples, ART adaptively inserts recognition categories to encode new input samples that are significantly distinct from existing prototypes. Therefore, ART2-C tends to generate reference prototypes using distinct samples, which in turn makes the output clusters to be more dissimilar to each other. This unique neuron initialization mechanism appeared to be effective in representing diverse data patterns in the input set. Our observations were supported by the t-test validations in Table 5.

Table 5: Statistical significance of our cross-method comparisons between k-means, SOM, and ART2-C on the Reuters-21578 corpus. ">>" and ">" (or "<<" and "<") denote the left-side value is greater (or smaller) than the right-side value at significance level 0.01 and 0.05 respectively.

Cluster number = 25					
	T	Cmp	Sep	Ec	El
ART2-C vs. SOM	<	>>	<<	>>	<
ART2-C vs. k-means	<<	>>	<	>>	<
Cluster number = 81					
	T	Cmp	Sep	Ec	El
ART2-C vs. SOM	<<	>>	<	>>	<<
ART2-C vs. k-means	<<	>>	<<	>>	<<

It is interesting that, since the two categories of quality measures are based on the same theoretical fundament (i.e. to encourage intra-cluster compactness and inter-cluster separation), the comparison results using the two categories of measures generally support each other. For an experiment that produces a low *cluster compactness* score, the *cluster entropy* score is accordingly low. For an experiment that produces a low *cluster separation* score, the *class entropy* score is generally low as well.

Besides the clustering quality of the trio, we are particularly interested in the learning efficiency of each method. k-means utilizes a relatively slow off-line learning paradigm in modifying cluster centroids to achieve minimal mean errors within clusters iteratively. SOM updates the winner's neighbors during the encoding of each input, which could be rather computationally expensive. The time cost of SOM's learning was generally comparable to that of k-means in our experiments. The learning paradigm of ART guarantees that only input data with significant similarity are to be encoded together. Therefore ART2-C is capable of encoding the inputs using a relatively high learning rate without causing oscillation. With this architectural advantage, ART2-C showed a promising learning efficiency in our experiments. The number of learning iterations used by ART-C was generally four to six times less than those used by SOM and k-means. The training time of ART-C was significantly less than those of SOM and k-means. This suggested the strength of ART-C in handling massive real-life data on the fly.

Based on the discussions above, we summarize our comparative findings below:

- While the clustering output of SOM showed a slightly worse quality than that of k-means in some of our experiments, the performances of k-means and SOM are roughly comparable.
- The intra-cluster compactness of ART2-C's output is generally worse than those of k-means and SOM's outputs.
- The inter-cluster separation of ART2-C's output is generally better than those of k-means and SOM's outputs.
- ART2-C shows a significantly higher efficiency than k-means and SOM.

6 Conclusions

Clustering is one of the major learning tasks in data mining research. The fundament of the clustering task is to group the input data into clusters such that data points in the same cluster are more similar to each other than to points in other clusters.

A clustering system consists of three major components, namely the feature representation, the definition of pattern proximity measure, and the grouping paradigm. This chapter gives insightful illustration of the various factors in each aspect that affect the output of a clustering system. Our review work concludes that cluster analysis and interpretation serves as an important feedback for selecting the parameters in each stage towards an optimal clustering output.

While the evaluation of the clustering result may be subjective, quantitative assessment of the clustering quality is of great value for various research areas. Since the objective of a clustering method is to maximize the intra-cluster compactness as well as the inter-cluster separation, a multitude of evaluation measures based on these two principles have been extensively studied in the literature. This chapter reviews and introduces two sets of evaluation measures, namely evaluation measures based on cluster distribution and evaluation measures based on class conformation. Our experiments showed the strength of these evaluation measures in various tasks, including discovering the inherent data distribution for suggesting the optimal number of clusters, choosing a suitable pattern proximity measure for a problem domain, and comparing various clustering methods for a better understanding of their learning characteristics.

References

[1] D. Boley. Principal direction divisive partitioning. *Data Mining and Knowledge Discovery*, 2(4):325–344, 1998.

[2] G.A. Carpenter, S. Grossberg, and D.B. Rosen. ART 2-A: An adaptive resonance algorithm for rapid category learning and recognition. *Neural Networks*, 4:493–504, 1991.

[3] G.A. Carpenter, S. Grossberg, and D.B. Rosen. Fuzzy ART: Fast stable learning and categorization of analog patterns by an adaptive resonance system. *Neural Networks*, 4:759–771, 1991.

[4] G.A. Carpenter and S. Grossberg. ART 2: Self-organization of stable category recognition codes for analog input patterns. *Applied Optics*, 26:4919–4930, 1987.

[5] G.A. Carpenter and S. Grossberg. A massively parallel architecture for a self-organizing neural pattern recognition machine. *Computer Vision, Graphics, and Image processing*, 34:54–115, 1987.

[6] D.L. Davis and D.W. Bouldin. A cluster separation measure. *IEEE Transactions on Pattern Analysis and Machine Intelligence*, 1(2), 1979.

[7] J.C. Dunn. Well separated clusters and optimal fuzzy-partitions. *Journal of Cybernetics*, 4:95–104, 1974.

[8] A. Flexer. On the use of self-organizing maps for clustering and visualization. In *Proceedings of the 3rd European Conference on Principles of Data Mining and Knowledge Discovery (PKDD)*, pages 80–88, 1999.

[9] E. Gokcay and J.C. Principe. A new clustering evaluation function using Renyi's information potential. In *International Conference on Acoustics, Speech, and Signal Processing (ICASSP)*, 2000.

[10] M. Halkidi and M. Vazirgiannis. A data set oriented approach for clustering algorithm selection. In *Proceedings of the 5th European Conference on Principles of Data Mining and Knowledge Discovery (PKDD)*, 2001.

[11] M. Halkidi, M. Vazirgiannis, and I. Batistakis. Quality scheme assessment in the clustering process. In *Proceedings of the 4th European Conference on Principles of Data Mining and Knowledge Discovery (PKDD)*, pages 265–276, 2000.

[12] J. He, A.H. Tan, and C.L. Tan. ART-C: A neural architecture for self-organization under constraints. In *Proceedings of International Joint Conference on Neural Networks (IJCNN)*, pages 2550–2555, 2002.

[13] A.K. Jain and R.C. Dubes. *Algorithms for Clustering Data.* Prentice-Hall, Upper Saddle River, NJ, 1988.

[14] A.K. Jain, M.N. Murty, and P.J. Flynn. Data clustering: A review. *ACM Computing Surveys*, 31(3):264–323, 1999.

[15] T. Kohonen. *Self-Organization and Associative Memory.* Springer-Verlag, Berlin, second edition, 1997.

[16] N.R. Pal and J. Biswas. Cluster validation using graph theoretic concepts. *Pattern Recognition*, 30(6), 1997.

[17] G.D. Ramkumar and Arun N. Swami. Clustering data without distance functions. *Bulletin of the IEEE Computer Society Technical Committee on Data Engineering*, 21(1):9–14, 1998.

[18] J.T. Tou and R.C. Gonzalez. *Pattern Recognition Principles.* Addison-Wesley, Massachusetts, 1974.

[19] D. Wettschereck and T.G. Dietterich. An experimental comparison of the nearest-neighbor and nearest-hyperrectangle algorithms. *Machine Learning*, 19(1):5–27, 1995.

[20] Y. Yang and X. Liu. A re-examination of text categorization methods. In *Proceedings of the 22nd Annual International ACM SIGIR Conference on Research and Development in Information Retrieval (SIGIR)*, pages 42–49, 1999.

[21] Y. Yang and J.P. Pedersen. A comparative study on feature selection in text categorization. In *Proceedings of the 14th International Conference on Machine Learning (ICML)*, pages 412–420, 1997.

[22] O.R. Zaiane, A. Foss, C.H. Lee, and W. Wang. On data clustering analysis: Scalability, constraints, and validation. In *Proceedings of the 6th Pacific-Asia Conference on Knowledge Discovery and Data Mining (PAKDD)*, pages 28–39, 2002.

CLUSTERING AND INFORMATION RETRIEVAL (pp. 135-159)
W. Wu, H. Xiong and S. Shekhar(Eds.)

Techniques for Textual Document Indexing and Retrieval via Knowledge Sources and Data Mining[1]

Wesley W. Chu
Computer Science Department
University of California, Los Angeles CA 90024
E-mail: wwc@cs.ucla.edu

Victor Zhenyu Liu
Computer Science Department
University of California, Los Angeles CA 90024
E-mail: vicliu@cs.ucla.edu

Wenlei Mao
Computer Science Department
University of California, Los Angeles CA 90024
E-mail: wenlei@cs.ucla.edu

Contents

[1]This research is supported in part by NIC/NIH Grant #4442511-33780

1 Introduction

Efficient document retrieval that answers a user query is achieved by indexing. The current technique uses word stems to index a document [1]. Such a technique suffers from the inability to match words in a query with their related words such as synonyms, hypernyms and hyponyms [2] in the documents. Therefore, there are recent attempts to index the document based on conceptual terms, however, the knowledge sources are usually incomplete. As a result, past research reveals that although using conceptual terms for document indexing can solve some of the problems, it cannot outperform the word-stem-based model [3, 4, 5, 6]. To remedy the deficiency of the knowledge sources, we propose a phrase-based indexing model where we parse a document into phrases based on the conceptual terms in domain specific knowledge sources, and calculate the similarity between two documents using both the similarity between the concepts and the common word stems in them. Including word stems in addition to concepts in document similarity evaluation compensates for the shortcoming of using concept terms alone caused by the incompleteness of the knowledge sources.

When seeking specific information regarding a particular topic, the user often has to pose a general query with concept terms. For example, not knowing that contact lenses are a treatment option for keratoconus, the user has to request for treatment options for keratoconus in the query. This results in low retrieval precision since documents are indexed by the specific terms. To remedy such a shortcoming, we propose to substitute the general concept terms in the query with the specific terms. The level of relevancy of a specific term in the resulting query is determined by its co-occurrence with the general concept term, which can be mined from the corpus. Based on the query, the knowledge sources can identify the irrelevant conceptual terms and prevent them from being included in the query. Since the expanded queries match better with relevant documents, the document retrieval performance is improved.

We shall first present the phrase-based indexing technique and the experimental results to show the performance improvements of phrase-based indexing over word-stem- and concept-based indexing methods. Next we shall present the knowledge-based query extension technique and present the performance improvement derived from the query extension. Finally, we present an implementation of integrating the two proposed techniques in a medical digital library for retrieving medical textual records and reports.

2 Indexing without domain knowledge

To facilitate discussion, we shall use the following sample query in this section: "22-year-old with hyperthermia, leukocytosis, increased intracranial pressure, and central herniation. Cerebral edema secondary to infection, diagnosis and treatment." The first part of the query is a brief description of the patient; the second part is the information need.

2.1 Word stem indexing

A document is commonly represented as a vector of terms in a *vector space model* (VSM)[1]. The basis of the vector space corresponds to distinct terms in a document collection. Components of the document vector are the weights of the corresponding terms that represent their relative importance in the document. In a naïve approach, we could treat a word as a term. Yet, morphological variants like "edema" and "edemas" are so closely related that they are usually conflated into a single *word stem*, e.g., "edem" by stemming [1, 7]. Our sample query thus consists of word stems "hypertherm," "leukocytos," "increas," "intracran," "pressur," etc. Word stems are usually treated as notational, rather than conceptual entities. Two word stems are considered unrelated if they are different. For example, the stem of "hyperthermia" and that of "fever" are usually considered unrelated despite their apparent relationship. In stem-based VSM, word stems constitute the basis of the vector space. The base vectors are orthogonal to each other because different word stems are considered unrelated. The weight $w^s_{\alpha,u}$ of a word stem u in a document α is determined by the number of times u appears in α (known as the *term frequency*) and the number of documents that contain u (known as the *document frequency*) following the TF-IDF (term frequency, inverse document frequency) scheme [1]. In essence, the more often u appears in α, the more important u is in α. On the other hand, the more documents u belongs to, the less disambiguating power it has, and thus the less important it is.

2.2 Multi-word indexing

Word stems are widely used as index terms. To improve retrieval accuracy, multi-word index terms may be used for indexing.

Several methods exist for defining multi-word indexing terms. An n-gram is defined as an ordered sequence of n words taken from a document. For example, "several methods" and "methods exist" are the first two bi-grams of the last sentence. Given a document d, by providing context lacking from isolated words, n-grams may more accurately model the content of documents.

Two other factors may influence the effectiveness of n-grams as indexing terms. First, n-grams depend on word order, thus "right upper lobe mass" is not equivalent to "mass right upper lobe." Second, n-grams are limited by word proximity, requiring that words appear next to one another in the original text. For example, in the text sample: "a mass is seen in the right upper lobe," here, "mass" and "right upper lobe" will only appear together if an 8-gram is used to model the text. Removing typical stop words from the sample results in "mass seen right upper lobe." Still, for the finding and anatomy descriptions to appear together requires an n-gram with a minimum length of 5 terms.

N-grams [8] may improve retrieval precision by providing additional context over isolated words. However, reliance on the original document's word order as well as word proximity may decrease retrieval recall and may require longer n-grams to be used to model documents.

We define an n-word combination as an unordered collection of n words taken from a document [9]. Given the text sample: "right upper lobe mass," there are 6 different 2-word combinations, including "right upper" and "upper mass." Unlike n-grams, n-word combinations (n-combos) do not depend on word order or proximity. Any set of n words can form an n-combo.

Removing the restriction on word order and proximity dramatically increases the number of potential n-combos. Given a document d of length l, there are $l!/(n!(l-n)!)$ n-word combinations in d. As the length of the documents grows, the number of n-combos grows dramatically (e.g., a 100 word document has the potential of 3,921,225 4-combos, a 200 word document has 64,684,950). Brute force calculation of all possible n-word combinations in a document, even for relatively small n, is too time and space expensive. In order to use n-word combinations, some method of limiting the search space must be defined. Furthermore, a method to select which n-combos should be used as indexing terms must be developed.

Although each document has a central theme (e.g., medical report describes an individual patient), the concepts useful for indexing are described in the individual sentences of the document. Limiting the search scope to individual sentences will dramatically decrease the time and space required to calculate n-word combinations, while focusing on relevant indexing terms. Furthermore, stop-word lists are employed to further reduce the search space by factoring out those words that do not carry any semantic significance.

New efficient algorithms based on pattern decomposition (PD) technique have been developed [10] to improve computation efficiency by an order of magnitude in finding the frequent item set as compared to the Aprior algorithm [11]. The PD algorithm makes the n-word combo indexing computation feasible.

When properly used, multiple-word combinations were shown to improve retrieval effectiveness for some special queries[9]. However, the retrieval effectiveness improvement for ad hoc queries is still questionable.

3 Indexing with domain knowledge

Using word stems to represent documents results in the inappropriate fragmentation of concepts such as "increased intracranial pressure" into its component stems "increas," "intracran," and "pressur." It is natural to replace word stems with concepts. Clearly, using concepts instead of single words or word stems as the vector space basis should produce a VSM that better mimics human thought processes, and therefore should result in more accurate retrieval. However, previous research showed not only no improvements, but degradation in retrieval accuracy when concepts were used in document retrieval [3, 4, 5, 6] except when documents were very short [12].

This is because using concepts is more complex than using word stems. First, concepts are usually represented by multi-word phrases such as "increased intracranial pressure." More importantly, there exist synonymous and polysemous phrases. Two phrases sharing a concept are *synonymous*, and phrases that could represent more than one concept are *polysemous* [2]. For example, "hyperthermia" and "fever" are synonymous because they share the same concept "an abnormal elevation of the body temperature."

At the same time, "hyperthermia" is polysemous, because in addition to the above meaning it also means "a treatment in which body tissue is exposed to high temperature to damage and kill cancer cells." Synonyms can be identified with the help of a dictionary or a thesaurus. Determining which concept a particular polysemous phrase represents is known as *word sense disambiguation* (WSD) [13]. Third, some concepts are related to one another. Hypernym and hyponym relations are important conceptual relations. If we say "an x is a (kind of) y" then concept x is a *hyponym* of concept y, and y is a *hypernym* of x [2]. For example, "hyperthermia" is a hyponym of "high body temperature;" and "high body temperature" is a hypernym of "hyperthermia."

Concept identifiers are usually used to identify concepts. Using UMLS [14] as a knowledge source, our sample query becomes (15967, 203597), (23518), and (151740) etc., representing "hyperthermia," "leukocytosis," and "increased intracranial pressure," etc., respectively.

In concept-based VSM, the basis of the vector space consists of distinct concepts. To model the relationship of such concepts as "hyperthermia" and "elevated body temperature" we remove the orthogonality constraint on base vectors. Base vectors for two related concepts form an acute angle. It is only when we cannot find any reasonable relations between two concepts that we treat their corresponding vectors as orthogonal. The cosine of the angle between two concept vectors is defined as the conceptual similarity between the corresponding concepts. The conceptual similarity thus ranges from 0 to 1 with 0 indicating unrelated and 1 indicating synonymous concepts.

To study the effects of conceptual similarities, we shall compare two cases. In one case, we assume all different concepts are unrelated. Therefore, all base vectors of the vector spaces are orthogonal to one another. In the other case, we derive conceptual similarities from knowledge sources. The resulting base vectors are no longer mutually orthogonal.

We derive the weight w^c_{α,x_i} of the ith concept x_i in a document α using a slightly modified version of the TF-IDF scheme. Higher weights are assigned to longer phrases that correspond to more specific concepts. For example, if the term frequencies and document frequencies for "increased intracranial pressure" and "hyperthermia" were identical, the former concept would obtain a higher weight than the latter.

3.1 Phrase-based VSM[15]

Conceptual similarities needed in concept-based VSM are derived from knowledge sources. The quality of such VSM therefore depends heavily on the quality of the knowledge sources. The absence of certain conceptual relations in the knowledge sources could potentially degrade retrieval accuracy. For example, treating "cerebral edema" and "cerebral lesion" as unrelated is potentially harmful. Noticing their common component word "cerebral" in the above phrases, we propose phrase-based VSM to remedy the incompleteness of the knowledge sources.

In phrase-based VSM, a document is represented as a set of phrases. Each phrase may correspond to multiple concepts (due to polysemy) and consist of several word stems. Our sample query now becomes [(15967, 203597), ("hypertherm")], [(23518), ("leukocytos")] and [(151740), ("increas", "intracran", "pressur")] etc.

Following the TF-IDF schemes in stem-based and concept-based VSMs, we can derive the stem weight $w^s_{\alpha,u_{i,k}}$ of the kth stem $u_{i,k}$ and the concept weight $w^c_{\alpha,x_{i,m}}$ of the mth concept $x_{i,m}$ in phrase i of α.

Similar to concept-based VSM, we study two cases. In one, different concepts are treated as unrelated; in the other, concepts may be related. In both cases, distinct word stems are assumed to be unrelated.

3.2 Document similarity

The similarity of two documents α and β is defined as the cosine of the angle between their corresponding document vectors $\vec{\alpha}$ and $\vec{\beta}$,

$$\mathrm{sim}(\alpha, \beta) = \cos(\vec{\alpha}, \vec{\beta}) = \frac{\vec{\alpha} \cdot \vec{\beta}}{\sqrt{\vec{\alpha} \cdot \vec{\alpha}}\sqrt{\vec{\beta} \cdot \vec{\beta}}}$$

To calculate phrase-based document similarity, we shall extend the document vector dot product $\vec{\alpha} \cdot \vec{\beta}$ and denote the *extended dot product* (EDP) as $\vec{\alpha} \circ \vec{\beta}$. Using the EDP in place of the dot product, we derive document similarity as,

$$\mathrm{sim}(\alpha, \beta) = \frac{\vec{\alpha} \circ \vec{\beta}}{\sqrt{\vec{\alpha} \circ \vec{\alpha}}\sqrt{\vec{\beta} \circ \vec{\beta}}} \tag{1}$$

EDP Derivation To derive the EDP in the phrase-based VSM, we first consider concepts without polysemy,

$$\vec{\alpha} \circ \vec{\beta} = \sum_{i,j} S^c_{i,j}$$

where $S^c_{i,j}$ is the conceptual contribution of phrase i in α and phrase j in β to the EDP. Assuming that each phrase represents a single concept, we have,

$$S^c_{i,j} = w^c_{\alpha,x_i} w^c_{\beta,y_j} s(x_i, y_j) \tag{2}$$

where $s(x_i, y_j)$ is the conceptual similarity between the ith concept x_i in α and the jth concept y_j in β. When different concepts are treated as unrelated, $s(x, y)$ in (2) is reduced to the Kronecker delta function,

$$\delta(x,y) = \begin{cases} 1 & x = y \\ 0 & x \neq y \end{cases}$$

When concepts may be related, we derive conceptual similarities from knowledge sources.

In order to use (2) in the presence of polysemy, we need to disambiguate senses. To avoid WSD cost, we use the most popular concept that a phrase represents as the phrase's meaning. Alternatively, we derive the conceptual contribution to the similarity between two phrases using an aggregation of (2) over all possible concept pairs. Each pair consists of one concept from each phrase.

The contribution of word stems to the EDP is the sum of the weight products for those word stems common to both phrases,

$$S^s_{i,j} = \sum_{k,l} w^s_{\alpha,u_{i,k}} w^s_{\beta,v_{j,l}} \delta(u_{i,k}, v_{j,l}) \tag{3}$$

where $u_{i,k}$ and $v_{j,l}$ are the kth word stem in phrase i in α and the lth word stem in phrase j in β, respectively.

Given the contribution of concepts (2) and stems (3), we select the larger of the two as the contribution of phrase i in α and phrase j in β to the EDP. Furthermore, we assign different similarity contribution factors f^c and f^s to the concept and stem contributions respectively, to indicate the relatively importance of each contribution. Thus,

$$\vec{\alpha} \circ \vec{\beta} = \sum_{i,j} \max(f^c S^c_{i,j}, f^s S^s_{i,j}) \tag{4}$$

Such selection remedies the incompleteness of the knowledge sources. $\vec{\alpha} \circ \vec{\alpha}$ and $\vec{\beta} \circ \vec{\beta}$ can be derived similar to (4). The document similarity can then be computed from (1) using these EDPs. Because we use an identical EDP formula to calculate the denominator and numerator in (1), the ratio of the similarity contribution factors f^c/f^s, instead of their absolute values, which is important. Changing the values of f^c and f^s while maintaining their ratio f^c/f^s produces identical document similarity $\text{sim}(a, b)$.

Conceptual Similarity in Hypernym Hierarchy Given a hypernym hierarchy, the conceptual similarity $s(x, y)$ between concepts x and y depends on both the distance between them in the hierarchy and their generality. When two concepts are farther apart in the hypernym hierarchy, they are less similar – a concept is less similar to its grandparent than to its parent in the hypernym hierarchy. Thus we define the conceptual similarity to be inversely proportional to the number of "hops" between x and y, $d(x, y)$. The generality of a concept x can be derived from the number of its descendants $D(x)$. The more descendants x has, the more general it is. A general concept like "disease" has many more descendants than a more specific concept like "hyperthermia" has. Because of the exponential growth of the number of descendants when a concept moves up a tree structure, we take the logarithm of the number of descendants in conceptual similarity calculation. The conceptual similarity is therefore defined to be inversely proportional to the logarithm of the number of descendants of the two. A final consideration is the boundary case when we reach the leaves of the hypernym tree. Let us assume we have two concepts x_0 and y_0, where x_0 is the only direct hypernym of y_0, y_0 is the only hyponym of x_0, and y_0 has no hyponym of its own. Concepts x_0 and y_0 are so much alike that we define the conceptual similarity between them to be c close to 1, say 0.9, to represent such closeness. As a result, the conceptual similarity between concepts x and y is,

$$s(x, y) = \frac{c}{d(x, y) \log_2(1 + D(x) + D(y))} \tag{5}$$

3.3 Experimental results

The Test Collection, OHSUMED OHSUMED [16] is a large test collection used in many information retrieval system evaluations. The test set consists of a reference collection, a query collection, and a set of relevance judgments.

The reference collection is a subset of the MEDLINE database. Each reference contains a title, an optional abstract, a set of MeSH headings, author information, publication type, source, a MEDLINE identifier, and a sequence identifier. The query collection consists of 106 queries. Each query contains a patient description, an information request, and a sequence identifier. The sample query we use in this paper is query 57 in the collection. 14,430 references out of the 348K are judged by human experts to be not relevant, possibly relevant, or definitely relevant to each query. We use the title, the abstract, and the MeSH headings to represent each document; the patient description and the information request represent each query.

The Knowledge Source UMLS [14] is a medical lexical knowledge source and a set of associated lexical programs. The knowledge source consists of UMLS Metathesaurus, SPECIALIST lexicon, and UMLS semantic network. Especially of interest to us is its central vocabulary component - the Metathesaurus. It contains biomedical phrases from more than 60 vocabularies and classifications. The Metathesaurus contains 1.6M phrases representing over 800K concepts.

A concept unique identifier (CUI) identifies each concept. Because of synonymy, multiple phrases can be associated with one CUI. For example, 71 phrases in 15 languages are associated with CUI 15967. Some English phrases for this CUI are "fever," "hyperthermia," "high body temperature," and "temperature, high." UMLS tends to assign a smaller CUI to the more popular sense of a phrase. For example, the CUI for the "high body temperature" sense of "hyperthermia" is 15967, while the CUI for its "treatment" sense is 203597. Therefore, we use the concept with the smallest CUI in conceptual contribution calculation (2). Our experimental results show that such heuristic produces retrieval accuracy comparable to that produced by the aggregation approach where we consider all conceptual similarities due to different sense combinations from the phrases.

The Metathesaurus encodes many conceptual relations. We concentrate on hypernym relations. Two pairs of relations in UMLS roughly correspond to the hypernym relations: the RB/RN (broader than/narrower than) and the PAR/CHD (parent/child) relations. For example, "hyperthermia" has a parent concept "body temperature change." The PAR and CHD relations are redundant. If (x, y) is in PAR, then (y, x) is in CHD; and vice versa. Similarly, the RB and RN relations are redundant. Therefore, we combine the 838K RB and 607K PAR relations into a single hypernym hierarchy.

Hypernymy is transitive [17]. For example, "sign and symptom" is a hypernym of "body temperature change" and "body temperature change" is a hypernym of "hyperthermia," so "sign and symptom" is also a hypernym of "hyperthermia." However UMLS Metathesaurus encodes only the direct hypernym relations but not the transitive closure. We derive the transitive closure of the hypernym relation and use (5) to calculate the conceptual similarities.

Phrase Detection Given a set of documents (106 queries and 14K judged documents of OHSUMED), we need to detect any occurrences in a set of phrases (1.3M phrases in UMLS). We adopt the Aho-Corasick algorithm [18] for the set-matching problem to detect phrases:

First, Aho-Corasick algorithm detects *all* occurrences of any phrase in a document. But we only keep the longest, most specific phrase. For example, although both "edema" and "cerebral edema" are detected in the sample query, we keep only the latter and ignore the former.

Second, to detect multi-word phrases, we match stems instead of words in a document with UMLS phrases. We use Lovins stemmer [7] to derive word stems. To avoid conflating different abbreviations into a single stem, we define the stem for a word shorter than four characters to be the original word.

Third, stop-word removal is performed after the multi-word phrase detection. In this way, we correctly detect "secondary to" and "carcinoma" from "cerebral edema secondary to carcinoma." We would incorrectly detect "secondary carcinoma" if the stop-words ("to" in this case) were removed before the phrase detection.

Discussion of Results To calculate retrieval accuracy using precision-recall [1], we combined the "possibly relevant" and "definitely relevant" judgments in OHSUMED into a single relevant category. Based on the type of VSM, we calculate the document similarity between each of the 14K documents and each of the 105 queries (one query does not have relevant document). For a given VSM and a query, we rank the documents from the most to the least similar to the query.

When a certain number of documents are retrieved, *precision* is the percentage of retrieved documents that are relevant; and *recall* is the percentage of the relevant documents that has been retrieved so far. We evaluate the retrieval accuracy by interpolating the precision values at eleven recall points. The overall effectiveness of different VSM is then compared by averaging over the performance of all the 105 queries (Figure 1). The average of the eleven precision values gives an overview of the effectiveness of each VSM.

1. The baseline (Stems) uses stem-based VSM. Its 11-point average precision is 0.375.

2. Considering the contribution of concepts only, and treating different concepts as unrelated (Concepts Unrelated), we arrive at an 11-point average precision of 0.270, which is a 28% decrease from the baseline.

3. Similar to 2, but taking the concept inter-relationship into consideration (Concepts), we achieve a significant improvement over 2. The

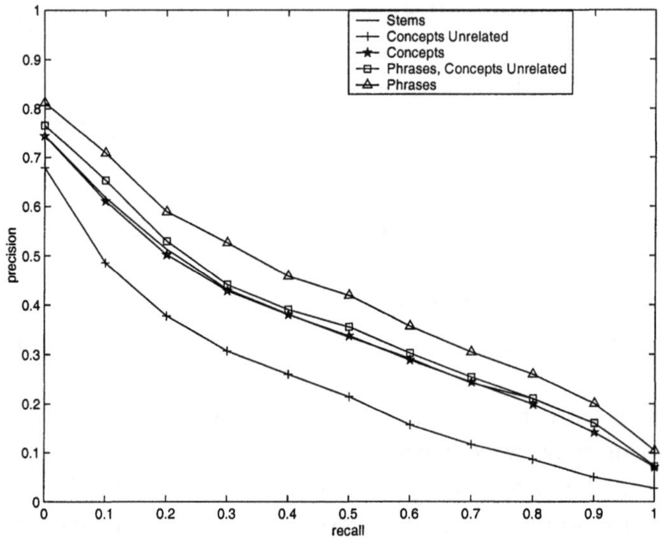

Figure 1: Comparison of the average precision-recall over 105 queries.

average accuracy is similar to that of the baseline.

4. Considering contribution of both concepts and word stems in a phrase, but treating different concepts as unrelated (Phrases, Concepts Unrelated), we arrive at an 11-point average precision of 0.386, a 3% improvement over the baseline. In both this case and the following case 5, we use equal similarity contribution factors $f^s = f^c$.

5. Similar to 4, but taking concept interrelations into consideration (Phrases), we achieve an 11-point average precision of 0.431, which is a significant 15% improvement over the baseline.

Our experiment results reveal that viewing documents as concepts only and treating different concepts as unrelated can cause the retrieval accuracy to deteriorate (case 2). Considering concept inter-relations (case 3) or relating different phrases by their shared word stems (case 4) can both improve retrieval accuracy. The extended dot product combines contributions from the concepts and word stems. The phrase-based VSM utilizes such extended dot products and yields significant improvement in retrieval accuracy. To study the relative importance of the concept contribution (2) and stem contribution (3) to the EDP, we vary the similarity contribution factors f^c and f^s in (4) and calculate the average of the 11-point average precisions over the 105 queries. Varying from the stem-only case $(f^s, f^c) = (10, 0)$ to the

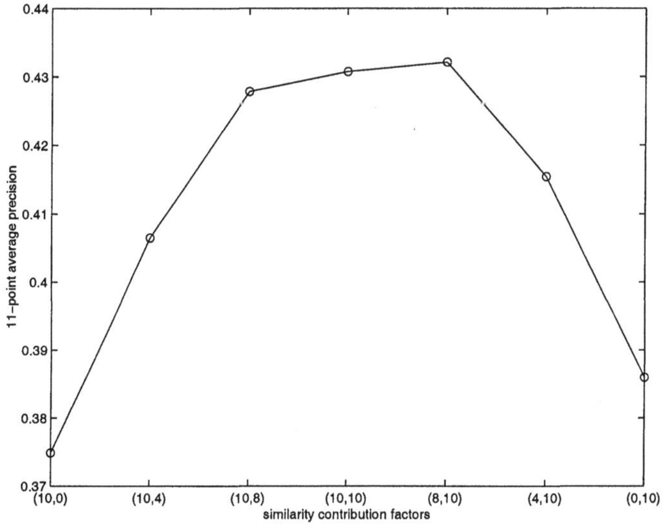

Figure 2: Effect of similarity contribution factors (f^s,f^c) on the average precision for phrase-based indexing.

concept-only case $(f^s, f^c) = (0, 10)$, we obtain the effect of the similarity contribution factors on the average precision for phrase-based VSM (Figure 2). Note that the average precision is relative constant in the range of $(f^s, f^c) = (10, 8)$ to $(f^s, f^c) = (8, 10)$. As a result, we use equal similarity contribution factor $f^s = f^c$ to obtain best average precision results. Thus (4) reduces to

$$\vec{\alpha} \circ \vec{\beta} = \sum_{i,j} \max(S^c_{i,j}, S^s_{i,j})$$

The 11-point average precision for the stem-only case $(f^s, f^c) = (10, 0)$ in Figure 2 is the average of the eleven precision values for case 1 (Stems) in Figure 1. Similarly, the concept-only case $(0, 10)$ in Figure 2 corresponds to case 3 (Concepts) in Figure 1; and the equal similarity contribution factor case $(10, 10)$ in Figure 2 corresponds to case 5 (Phrases) in Figure 1.

4 Improving retrieval performance via query expansion

In real application, a user's query often consists of two parts: a key concept, x_{key}, which is the user's main objective in mind, and general supporting concepts, x_s, which specify certain aspects of x_{key}. For example, in a query

"Keratoconus, treatment options", "Keratoconus" (an eye disease) is the key concept whereas "treatment options" is a general supporting concept. This kind of query formation is typical in the 106 OHSUMED queries [16], some of which are shown in Table 1. A recent study [19, 20] also reveals that doctors' clinical questions about patient care can be generalized into a limited number of categories. Seven out of the top ten categories (a 61.68% coverage over all questions studied) used this formation of a key concept plus general supporting concepts. In the medical domain, general supporting concepts form a limited vocabulary, e.g. "treatment options", "work up", "therapy", "diagnosis" and "epiology" in OHSUMED, and can provided by domain experts.

Although such techniques are easy to form queries, the queries do not match well with relevant documents that use specific supporting concepts. In the example above, documents discussing treatment options for keratoconus use specific terms such as "contact lens" or "penetrating keratoplasty" instead of the general terms of "treatment" or "option". To remedy this problem, we propose to substitute the general supporting concepts with specific concepts used in the relevant documents. The rewritten query matches better with the relevant documents, thus improving the retrieval effectiveness. In this section, we will first review the past research on query expansion. Second, we will propose a knowledge-based approach that leads to more focused query expansion than traditional methods. The weight assigning scheme is discussed in 4.3. Finally we will present experimental results comparing the knowledge-based query expansion with basic expansion approaches.

4.1 Previous research on automatic query expansion

Using automatic query expansion to solve the query-document mismatch problem has been studied for decades [21, 22, 23, 24, 25, 26]. The basic idea is to first compute all terms that highly co-occur with the original query terms, and then append them to the original query. Co-occurrence measures are obtained automatically from the corpus from which documents are retrieved. It is assumed that terms highly co-occurring with the original query terms are likely to be synonyms or semantically related. Therefore, adding such extra terms is supposed to help retrieve relevant documents that do not use the exact terms in the original query. However, in early works by Sparck Jones [21, 22], the effectiveness of query expansion was unable to be justified. Not until the early nineties when large corpuses became common in research, and when the vector model and different vector weighting schemes had been intensively studied [1, 27], were researchers able

Table 1: Sample OHSUMED queries which each contain a key concept and general supporting concepts. Key concepts are shown in capital letters and general supporting concepts are in italics

Query ID	*Original Query Form*
13	**LACTASE DEFICIENCY** *therapy options*
16	**CHRONIC FATIGUE SYNDROME**, *management and treatment*
37	**FIBROMYALGIA / FIBROSITIS**, *diagnosis and treatment*
38	**DIABETIC GASTROPARESIS**, *treatment*
42	**KERATOCONUS**, *treatment options*
43	**BACK PAIN**, *information on diagnosis and treatment*
47	**URINARY RETENTION**, *differential diagnosis*
53	**LUPUS NEPHRITIS**, *diagnosis and management*

to show consistently improved retrieval effectiveness via query expansion [23, 24, 25]. In [23], word stems are used to index queries and documents. A stem is represented by a vector of documents in which each component is weighted using the normalized $tf * idf$ measure [27]. Using VSM, the co-occurrence measure between two stems is computed. The weight for an expanded stem is the average co-occurrence measure between that stem with each respective original query stem. Considering term co-occurrence on an entire document basis may be too coarse. To remedy this problem, [24] proposed an approach in which two terms are considered as co-occurring only when they appear close enough to each other (e.g. in one paragraph or in the distance of a few sentences). Using the same dataset as [24], this method obtained better retrieval accuracy. Combining this idea of co-occurrence computation with the blind feedback method further improved retrieval performance [25].

4.2 A knowledge-based approach to selectively expand concepts that are relevant to the original query

Most of the previous studies used a stem-based approach. In other words, queries are indexed and expanded by word stems. In these studies, stems

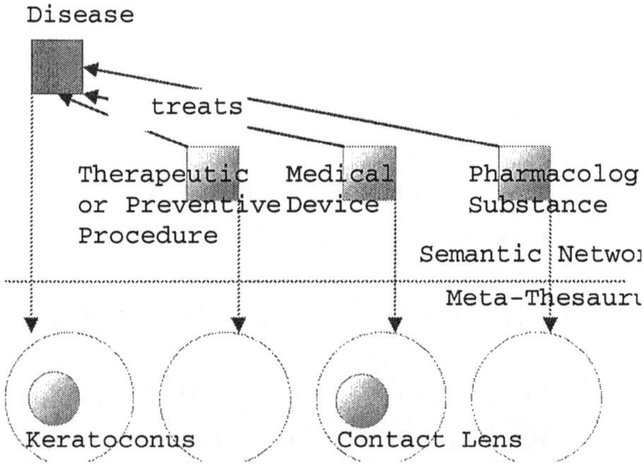

Figure 3: Mapping from concepts to semantic types and the relationships among semantic types in UMLS

are not associated with any domain knowledge. As a result, it is impossible to distinguish stems that are truly semantically related to the original query from those that are not, even though they all highly co-occur with the original query stems. Also, it is hard to comprehend stems that are expanded into the query, making it even harder to refine the expansion operation. For the sample query of "Keratoconus, treatment options", we computed 10 word stems that are expanded into the query with the highest weights (Table 2). The experimental settings were the same as [23]. As can be seen from the table, not all components added to the query are related to "treatment options", resulting in a less focused expansion. Therefore we propose a knowledge-based approach to obtain more focused query expansion. For our study, we used UMLS as the knowledge source and OHSUMED as the corpus. UMLS consists of three parts, the Metathesaurus, the Semantic Network and the SPECIALIST Lexicon. A group of concepts in the Metathesaurus are abstracted into one semantic type in the Semantic Network. Semantic types are associated with each other by a limited number of relationships. Figure 3 shows that "Keratoconus" and "Contact lens" belong to the semantic types of "Disease or Syndrome" and "Therapeutic or Preventive Procedure", respectively; "Procedure" is one of the types that "treats" "Disease". However, whether there is a "treats" link between "Keratoconus" and "Contact Lens" is not indicated by UMLS.

In the knowledge-based query expansion approach, we first assume that the key concept section of the query can be marked out by the user, and

Table 2: Top 10 stems that are expanded into the sample query "Keratoconus, treatment options"

Ranking based on weights assigned	*Word stem*
1	corne
2	keratoplast
3	acu
4	visu
5	epikeratoplast
6	lens
7	keratometer
8	penetr
9	len
10	ey

also the general supporting concepts are represented in terms of relationships among semantic types. Given the previous study on clinical question commonalities, and the improved completeness coverage of UMLS from version to version, this is a reasonable assumption. Second, the concept unique identifier (CUI) of the key concept can be automatically detected from the marked out section. Third, we abstract the CUI into its corresponding semantic type (e.g. "Disease or Syndrome" for "Keratoconus"), and follow the relationships as indicated by the general supporting concepts to a set of relevant semantic types. Finally, only concepts that belong to these relevant semantic types are considered as relevant specific supporting concepts and are added to the query. We also add the key concept's parents, children and siblings (all children of the parents of the key concept) to match relevant documents that use a broader or narrower topic. The scope is restricted to be one level up and down from the key concept, mainly because experimental results show that adding more ancestors or descendants hardly affects retrieval precision. Table 3 shows the top 10 concepts expanded into the sample query "Keratoconus, treatment options" using this knowledge-based approach. Compared to table 2, we note the knowledge-based query expansion is much more focused on the treatment aspect. In addition, the knowledge-based approach is able to filter out irrelevant specific supporting concepts. As a result, the expansion size is much smaller, which greatly reduced the computation complexity.

Table 3: Top 10 concepts that are expanded into the sample query using the knowledge-based approach

Ranking based on weights assigned	*Concepts*
1	Cornea Transplantation
2	Contact lens
3	Penetrating keratoplasty
4	Epikeratoplasty
5	Epikeratophakia
6	Eyeglasses
7	Buttons
8	Radial Keratotomy
9	Trephines
10	Thermokeratoplasty

4.3 Weight assignment for expanded vector components

For a specific supporting concept x, its weight should represent the degree of correlation between x and the key concept term, x_{key}. For example, "contact lens" is a treatment option for "Keratoconus" but not "Back pain", and therefore it should assign a larger weight in the expansion of "Keratoconus, treatment options" than that of "Back pain, treatment options". Let us now present a scalable method for such weight assignment. We shall first represent concepts into inverted document vectors, and then use the similarity between the two inverted document vectors to represent the correlation between the two concepts. Given a corpus of n documents, the inverted document vector for concept x, $\vec{\theta_x}$, is defined as an n-dimensional vector. Each component in this vector corresponds to a document, the weight of which represents the term frequency of concept x in that particular document. For example, if a corpus contains documents D_1, D_2 and D_3, and concept x occurs three, zero, and two times in these documents, respectively, then $\vec{\theta_x} =< 3, 0, 2 >$.

We further define the correlation between concepts x and y as:

$$\text{correl}(x, y) = \cos(\vec{\theta_x}, \vec{\theta_y}) = \frac{\vec{\theta_x} \cdot \vec{\theta_y}}{\sqrt{\vec{\theta_x} \cdot \vec{\theta_x}}\sqrt{\vec{\theta_y} \cdot \vec{\theta_y}}} \tag{6}$$

The correlation between two concepts ranged from 0 to 1. For example, if $\vec{\theta_x} =< 3, 0, 2 >, \vec{\theta_y} =< 6, 0, 4 >$, and $\vec{\theta_z} =< 0, 1, 0 >$, then $correl(x, y) = 1$

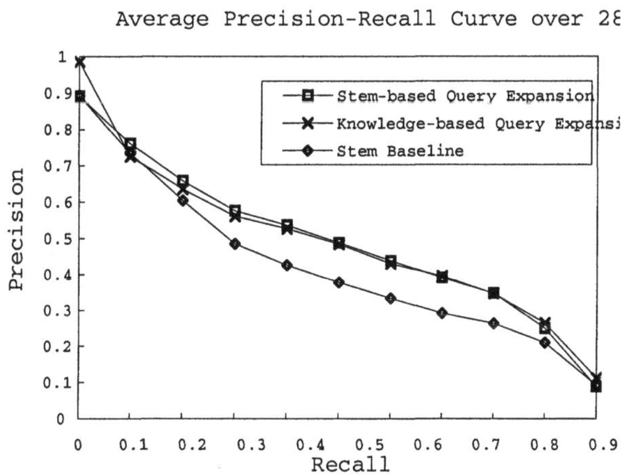

Figure 4: Retrieval performance improvements with knowledge-based query expansion

and $correl(x, z) = 0$. The correlation between all pairs of concepts can be computed offline and stored in a concept correlation table. For query expansion, the weight assigned to x is the correlation between a supporting concept x and x_{key}.

4.4 Experimental results

To evaluate the performance improvement of knowledge-based query expansion, we select 28 OHSUMED queries [16] that contain general supporting concepts as treatment and diagnosis options (see Table 1 for examples). We followed the procedure in [23] to generate the result for stem-based query expansion. We used the same stem-based VSM as in 3.4 to generate the baseline.

Our experimental results for the query set reveal the average query expansion size using traditional stem-based expansion is 14584 terms per query, while using UMLS, the average expansion size reduced to 235 terms per query. This represents more than an order of magnitude reduction in query expansion size.

The retrieval performance comparison for the set of selected OHSUMED queries is shown in Figure 4. We note that both expansion approaches obtain higher precision than the baseline, and the knowledge-based approach performed better than the stem-based approach in the low recall region (below 10% recall).

Knowledge-based query expansion using concept-based VSM can significantly improve the computation complexity (more than order of magnitude) over the full stem-based expansion. However, due to the insufficient knowledge granularity, the retrieval performance is not uniformly better than the traditional stem-based approach, particularly in the high recall region. Therefore, we are currently evaluating the knowledge-based query expansion using stem-based VSM. Our preliminary results reveal that knowledge-based query expansion with stem-based VSM has uniformly better retrieval performance than that of the traditional stem-based query expansion, but with less significant complexity reduction (an order magnitude of reduction). Thus, the user is provided with a choice of query expansion methods with different levels of computation and performance tradeoffs. We are also planning to investigate the knowledge-based query expansion using phrase-based VSM which will yield comparable performance as that of the knowledge-based expansion using stem-based VSM but with lower computation complexity.

5 Applications

We shall now apply the above techniques to a document retrieval system in a medical digital library. The system (Figure 5) consists of three subsystems: a Document Index Generator (DIG), a Query Expansion Processor (QEP) and a Document Retrieval Processor (DRP).

In the DIG, the Phrase Detector parses the document into phrases. Based on the concepts defined in the UMLS Metathesaurus, indices (both word stems and concepts) are generated for all the documents in the corpus. The Phrase Weight Calculator then computes the weights for all the phrases based on the corresponding term frequency and their inverse document frequency. The set of phrases and the associated weights transform the original corpus into a phrase-indexed corpus for phrase-based retrieval.

The concept correlation of concept x and y, $correl(x, y)$ (6), can be derived from the Phrase-indexed Corpus. Further based on the UMLS Metathesaurus Hyponym Hierarchy, the conceptual similarity of concept x and y, $s(x, y)$, can be computed (5) and later used to evaluate the query-document similarity (1). All of these operations can be done offline.

When a query is input into the QEP, the Phrase Detector and the General Concepts Detector parse the query and check whether there are any general supporting concepts. The General Supporting Concepts List is provided by domain experts. The detected general supporting concepts will be substituted with a set of specific supporting terms by the Query Expander

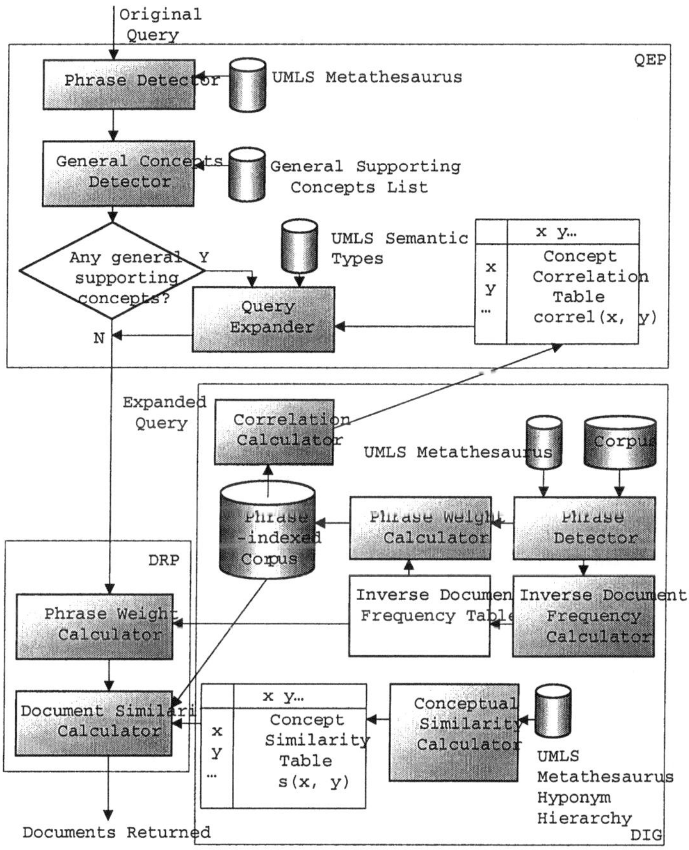

Figure 5: A phrase based indexing and query expansion document retrieval system

that refers to the appropriate UMLS Semantic Types and the Concept Correlation Table.

The DRP, based on the phrase-indexed query, retrieve a set of documents from the Phrase-indexed Corpus that are similar to the query conditions. The documents are ranked according to the phrase-based similarity measure.

6 Summary

We have presented indexing techniques for the retrieval of textual documents. First, we presented indexing techniques without domain knowledge, such as word stem and multi-word indexing and their shortcomings. Next we discussed indexing with domain knowledge of the corpus and developed a new vector space model that uses phrases to represent documents. Each phrase consists of multiple concepts and words. Similarity between two phrases is jointly determined by the conceptual similarity and their common word stems. Our experimental result reveals that the phrase-based VSM yields a 15% increase of retrieval effectiveness over that of the stem-based VSM. This improvement is because multi-word concepts are natural units of information, and using word stems in phrase-based document similarity compensates for the inaccuracy in conceptual similarities derived from incomplete knowledge sources.

We introduced a knowledge-based technique to rewrite a user query containing general conceptual terms into one containing specific terms. These specific supporting concepts are selected via knowledge sources and are related to the general supporting concept and the query's key concept. Weights for those specific concepts are assigned from data-mining the corpus. Experimental results show that retrieval using such expanded queries is more effective than the original queries. The average size of the expanded queries in the knowledge-based approach is much smaller (reduced by more than orders of magnitude) than that produced by the stem-based query expansion, and also yield better retrieval performance in the low recall region which is of interest to most applications. We also presented an implementation that integrates the above techniques into a digital medical library at UCLA for the retrieval of patient records, laboratory reports and medical literature.

References

[1] G. Salton and M.J. McGill, *Introduction to Modern Information Retrieval*, (McGraw-Hill Inc., 1983).

[2] G.A. Miller, R. Beckwith, C. Fellbaum, D. Gross and K. Miller, Introduction to WordNet: an On-line Lexical Database, *WordNet: an Electronic Lexical Database*, (1998), pp. 1-19.

[3] M. Mitra, C. Buckley, A. Singhal and C. Cardie, An Analysis of Statistical and Syntactic Phrases, *Proceedings of the Fifth RIAO Conference*, (1997), pp. 200-214.

[4] R. Richardson and A.F. Smeaton, Using WordNet in a Knowledge-based Approach to Information Retrieval, *Proceedings of the 17th BCS-IRSG Colloquium on Information Retrieval*, (1995).

[5] M. Sussna, Text Retrieval using Inference in Semantic Matanetworks, *PhD Thesis, University of California, San Diego*, (1997).

[6] E.M. Voorhees, Using WordNet to Disambiguate Word Sense for Text Retrieval, *In Proceedings of the 16th Annual ACM SIGIR Conferenceon-Research and Development in Information Retrieval*, (1993), pp. 171-180.

[7] J.B. Lovins, Development of a Stemming Algorithm. *In Mechanical Translation and Computational Linguistics*, 11(1-2), (1968), pp. 11-31.

[8] L.P. Jones, Jr. E.W. Gassie and S. Radhakrishnan, INDEX: The statistical basis for an automatic conceptual phrase-indexing system. *In Journal of American Society for Information Science*, 41(2) (1990), pp. 87-97.

[9] D. Johnson, W.W. Chu, J.D. Dionisio, R.K. Taira and H. Kangarloo, Creating and Indexing Teaching Files from Free-text Patient Reports. *In AMIA'99*, (1999).

[10] Q. Zou, W.W. Chu, D.B. Johnson and H. Chiu. Pattern decomposition algorithm for mining frequent patterns. *In Journal of Knowledge an Information System*, 4(4), (2002).

[11] R. Agrawal and R. Srikant. Fast algorithms for mining association rules. *In VLDB'94*, (1994), pp. 487-499.

[12] A.F. Smeaton and I. Quigley. Experiments on using Semantic Distances Between Words in Image Caption Retrieval. *In 19th Proc. ACM-SIGIR*, (1996), pp. 174-180.

[13] N. Ide and J. Véronis. Word Sense Disambiguation: the State of the Art. *In Computational Linguistics*, 24(1), (1998), pp. 1-40.

[14] National Library of Medicine. UMLS Knowledge Sources, 12th edition, (2001).

[15] W. Mao and W.W. Chu. Free text medical document retrieval via phrased-based vector space model. *Proc. AMIA'02*, (2002).

[16] W. Hersh, C. Buckley, T.J. Leone and D. Hickam. OHSUMED: an Interactive Retrieval Evaluation and New Large Test Collection for Research. *In Proc. 22nd ACM-SIGIR Conf.*, (1994), pp. 191-197.

[17] J. Lyons. Semantics, (1977).

[18] A.V. Aho and M.J. Corasick. Efficient String Matching: an Aid to Bibliographic Search. *In CACM*, 18(6), (1975), pp. 330-340.

[19] J.W. Ely, J.A. Osheroff, M.H. Ebell, G.R. Bergus, et al. 1999. Analysis of questions asked by family doctors regarding patient care. *British Medical Journal*, 319:358-361, 1999

[20] J.W. Ely, J.A. Osheroff, P.N. Gorman, M.H. Ebell, et al. 2000. A taxonomy of generic clinical questions: classification study. *British Medical Journal*, 321:429-432, 2000

[21] K. Sparck Jones. Automatic keyword classification for information retrieval. Butterworth, London, 1971

[22] K. Sparck Jones. Collecting properties influencing automatic term classification. *Information Storage and Retrieval*, 9:499-513, 1973

[23] Y. Qiu and H.P. Frei. Concept-based query expansion. In *Proc. 16th ACM-SIGIR*, pages 160-169, 1993

[24] Y. Jing and W.B. Croft. An association thesaurus for information retrieval. In *Proc. RIAO'94*, pages 146-160, 1994

[25] J. Xu and W.B. Croft. Query expansion using local and global document analysis. In *Proc. 19th ACM-SIGIR*, pages 4-11, 1996

[26] E.N. Efthimiadis. Query expansion. In *Annual Review of Information Science and Technology*, 31:121-187, 1996

[27] G. Salton and C. Buckley, 1988. Term weighting approaches in automatic text retrieval, *Information Processing & Management*, 24(5):513-523, 1988

Clustering and Information Retrieval (pp. 161-193)
W. Wu, H. Xiong and S. Shekhar(Eds.)

Document Clustering, Visualization, and Retrieval via Link Mining

Steven Noel
Center for Secure Information Systems
George Mason University, Fairfax, VA 22030, USA
E-mail: snoel@gmu.edu

Vijay Raghavan
Center for Advanced Computer Studies
The University of Louisiana at Lafayette, Lafayette, LA 70504, USA
E-mail: vraghavan@cacs.louisiana.edu

C.-H. Henry Chu
Center for Advanced Computer Studies
The University of Louisiana at Lafayette, Lafayette, LA 70504, USA
E-mail: cice@cacs.louisiana.edu

Contents

1 Introduction

Clustering for document retrieval has traditionally been done through word-based similarity. But this approach suffers from the ambiguity problems inherent in natural languages. Language-based processing can be augmented by analysis of links among document sets, i.e. hypertext Web links or literature citations. Indeed, early workers in information science recognized the shortcomings with word-based document processing. This led to the introduction of document processing based on literature citations [6]. An important development was the notion of co-citation [13], in which a document pair is associated by being jointly cited (or co-cited) by other documents. In general, clustering based on co-citation as a similarity measure is known to correspond well to document semantics. Spurred by the popularity of the Web, more recent approaches have been developed for analyzing hyperlinks, though primarily for search engine page ranking [12, 8].

Another important recent development in information science is association mining [1]. This measures how strongly certain sets of objects are associated through joint references. Only recently has it been recognized that association mining strength is a generalization of the classical co-citation similarity [11]. In particular, association mining generalizes relationships between pairs of documents to relationships among document sets of arbitrary cardinality. These higher-order associations capture relationships that are generally missed through pairwise co-citation similarities.

Association mining shares a particular aspect of the fundamental clustering hypothesis. Specifically, strongly associated sets are assumed to be composed of highly similar elements. But contrary to the clustering hypothesis, association mining lacks the assumption that objects in strongly associated sets are highly dissimilar to objects outside the set. Unlike the disjoint sets found in clustering, the sets in association mining are overlapping. This can lead to overwhelming numbers of combinations to consider, even for moderate-sized document collections. The challenge is to include higher-order associations while keeping the complexity manageable.

This chapter describes how link association mining can be applied to document clustering. It addresses the fundamental differences between strongly

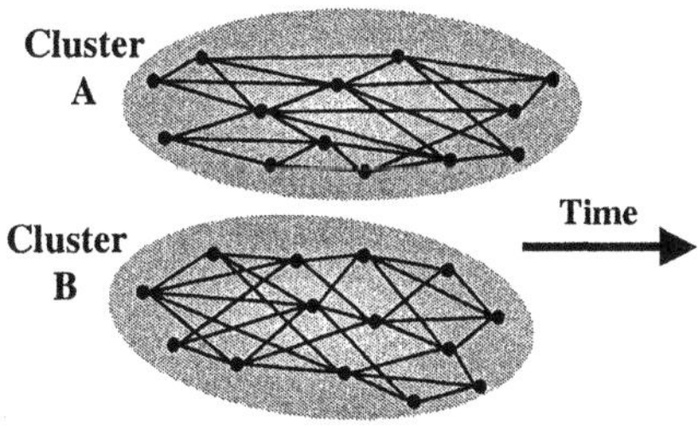

Figure 1: Link-based document clusters.

associated sets and document clusters, as well as how to meet the challenge of retaining higher-order information while maintaining low complexity. The approach described here includes higher-order link association information as similarity features for pairs of documents. From the pairwise document similarities, standard clustering approaches are then applied. In particular, it applies higher-order association similarities to cluster visualization for information retrieval tasks.

2 Link-Based Document Clustering

Traditionally, link-based document analysis has applied co-citation as a similarity measure for clustering. The typical goal was to discover subsets of large document collections that correspond to individual fields of study. Conceptually, documents form clusters if they share links among them in a certain sense, e.g. as shown in Figure 1.

Collections of linked documents (e.g. through citations or hyperlinks) can be modeled as directed graphs, as in Figure 2. A graph edge from one document to another indicates a link from the first to the second. In a matrix formulation, a binary adjacency matrix is formed corresponding to the document link graph. Assume that adjacency matrix rows are for citing (linking-from) documents and columns are for cited (linking-to) documents. Thus for adjacency matrix A, element $a_{i,j} = 1$ indicates that document i cites (or links to) document j, and $a_{i,j} = 0$ is the lack of citation (link).

Co-citation between a pair of documents is the joint citing (or hypertext linking) of the pair by another document, as shown in Figure 3. A traditional measure of similarity between a pair of documents is the number of documents that co-cite the pair, known as citation count. Taken over all pairs of documents, the co-citation count similarity serves as a compact

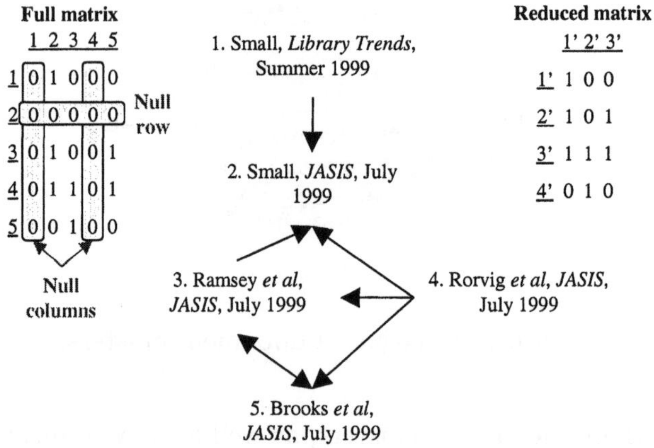

Figure 2: Adjacency matrix for document link graph.

representation of citation graph structure.

Co-citation between a pair of documents is the joint citing (or hypertext linking) of the pair by another document, as shown in Figure 3. A measure of similarity between a pair of documents is the number of documents that co-cite the pair, known as citation count. Taken over all pairs of documents, the co-citation count similarity serves as a compact representation of citation graph structure.

In terms of the document citation adjacency matrix A, co-citation count is a scalar quantity computed for pairs of matrix columns (cited documents). For columns j and k, co-citation count $c_{j,k}$ is then

$$c_{j,k} = \sum_i a_{i,j} a_{i,k} = a_j \cdot a_k = A^T A. \tag{1}$$

Here a_j and a_k are column vectors of A, i indexes rows, A^T is the transpose of A, and $x \cdot y$ is the vector dot (inner) product. Note that the product $a_{i,j} a_{i,k}$ represents single co-citation occurrences, which the summation counts. The co-citation count $c_{j,j}$ of a document with itself is simply a citation count, i.e. the number of times the document has been cited.

It is convenient to normalize the co-citation count $c_{j,k}$ through the linear transformation

$$\hat{c}_{j,k} = \frac{c_{j,k} - \min(c_{j,k})}{\max(c_{j,k}) - \min(c_{j,k})}, \tag{2}$$

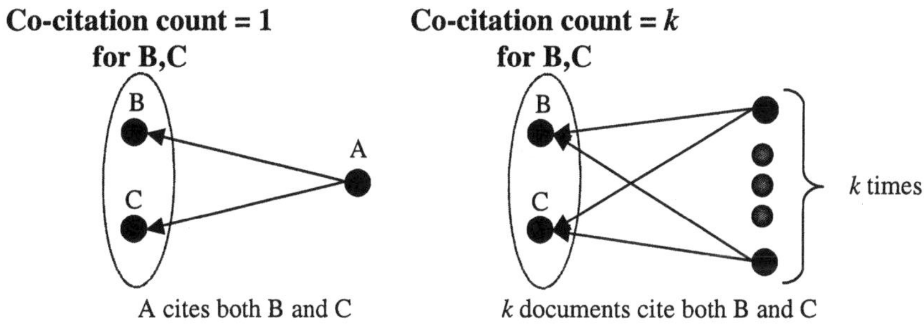

Figure 3: Co-citation document similarity.

yielding the normalized count $\hat{c}_{j,k} \epsilon [0,1]$. Here $\min(\cdot)$ and $\max(\cdot)$ are the minimum and maximum functions, respectively. Standard clustering and minimum spanning tree algorithms assume dissimilarities rather than similarities. We convert similarities to dissimilarities (distances) through the linear transformation

$$d_{j,k} = 1 - \hat{c}_{j,k}. \tag{3}$$

This results in distance $d_{j,k}$ between documents j and k, normalized to $d_{j,k} \epsilon [0,1]$.

Classical co-citation analysis relies on simple single-linkage clustering [16], because of its lower computational complexity given the typically large document collections. But a known problem with single-linkage clustering is a possible "chaining" effect, in which unrelated documents get clustered together through a chain of intermediate documents [14]. Figure 4 shows an example of single-linkage chaining, in which two clusters merge through a single pair of co-cited documents.

Alternative clustering criteria exist that are stronger than the single linkage criterion, e.g. average linkage and complete linkage (see Figure 5). These criteria are applied in an agglomerative clustering heuristic in which clusters that have the closest distance between them are iteratively merged. For single-linkage, the measure of distance between two clusters is the closest possible distance between objects in separate clusters. For average-linkage, cluster distance is the average of distances between objects in separate clusters. For complete-linkage, cluster distance is the furthest distance between objects in separate clusters. Thus single-linkage, average-

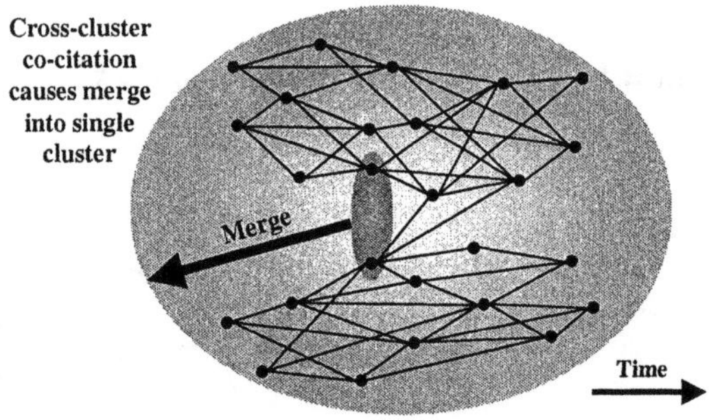

Figure 4: Chaining in co-citation single-linkage clustering.

linkage, and complete-linkage correspond to weak, intermediate, and strong clustering criteria, respectively.

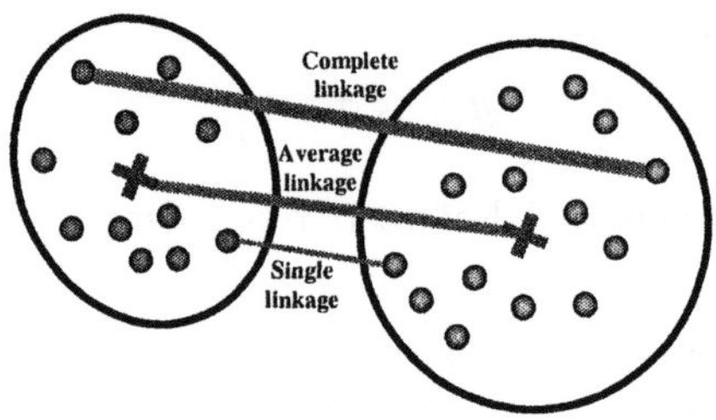

Figure 5: Inter-cluster distances for single-linkage, average-linkage, and complete-linkage.

3 Incorporating Higher-Order Link Information

Citation analysis has traditionally applied single-linkage clustering, because of its lower computational complexity. But being a weak clustering criterion, single-linkage has problems unless the data are inherently well clustered. Given the improved performance of modern computers, it becomes feasible to apply stronger clustering criteria in citation analysis. In fact, we apply a particularly strong criterion taken from the area of association mining.

This involves essentially higher-order co-citations, i.e. co-citations among document sets of arbitrary cardinality.

Beyond providing a stronger clustering criterion, another benefit of higher-order co- citations is with regard to user-oriented clustering. Here the user provides iterative feedback to help guide the clustering process, based on knowledge of the application domain. With pairwise distances, users can orient clustering by weighting distances for various document pairs, applying heavier weights to pairs whose similarities are more important. With higher-order similarities, this orientation can be generalized to weighting document sets of arbitrary cardinality.

3.1 Document Distances from Link Association Mining

Figure 6 illustrates the weak single-linkage criterion, and how it relates to citations. For the three cited documents in the example, there are three possible co-citation similarities (pairs of documents). As the example shows, only two of these similarities need to exceed the clustering threshold for the three documents to be considered a cluster, as long as they share a common document between the two pairs.

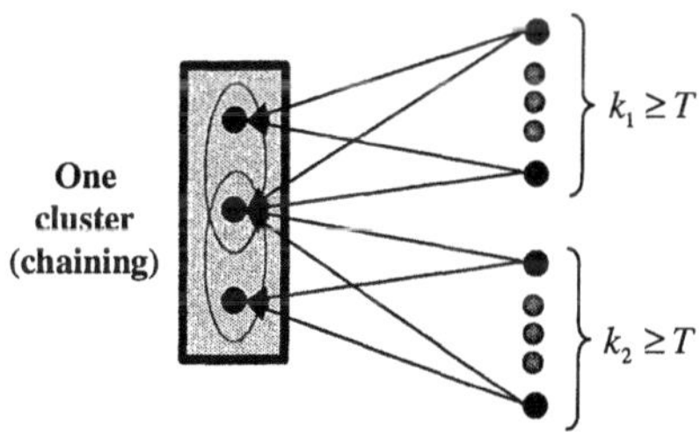

Figure 6: Single-linkage chaining with co-citation similarities.

In contrast, for the stronger clustering criterion of complete linkage all similarities for the three pairs need to exceed the threshold before the documents constitute a single cluster. This is shown in Figure 7. But notice that for this example, there is not even one document that cites all three of the clustered documents simultaneously. The complete- linkage criterion is a necessary but not sufficient condition for the simultaneous citing of all documents in a cluster.

But consider a generalization of co-citation similarity in which sets of arbitrary cardinality are considered for co-citation, as shown in Figure 8.

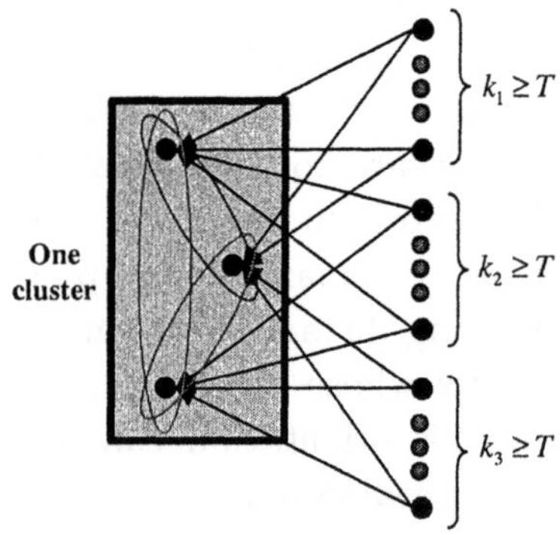

Figure 7: Stronger complete-linkage criterion with co-citation similarities.

That is, we define similarity as the number of times all the members of the set are simultaneously cited. Because the similarity involves more than two documents, it is higher order than pairwise similarity.

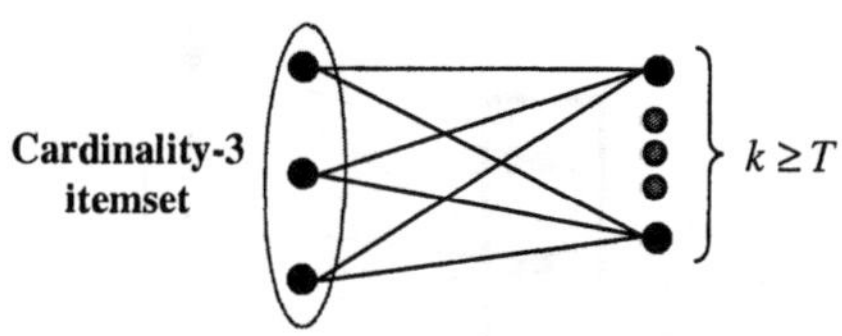

Figure 8: Itemset is an even stronger association than complete-linkage cluster.

We can specify a threshold value for these higher-order similarities (itemset supports) to identify sets whose similarities are sufficiently large. For our example, the only way the three cited documents could be considered a sufficiently similar set is if all three of them are cited more than the threshold number of times. In association mining, such a jointly referenced set is known as an *itemset*. The number of times that the set is jointly referenced (co-cited) is known as the itemset *support*. Itemsets whose members are sufficiently similar (have sufficient support) are known as *frequent* itemsets.

The extension from pairs of documents to sets of arbitrary cardinality means there is itemset overlap, that is, itemsets are non-disjoint. Such overlap is not possible with pairs of documents. Itemset supports of arbitrary cardinality are thus represented as *lattices* rather than $n \times n$ matrices for n documents.

In particular, itemsets are represented by the lattice of all subsets of a document collection. The subsets form a partial ordering, under the ordering relation of set inclusion. This is illustrated in Figure 9, via the Hasse diagram for visualizing partial orderings. The diagram shows the itemset lattice (excluding singletons and the empty set) for a set of four documents.

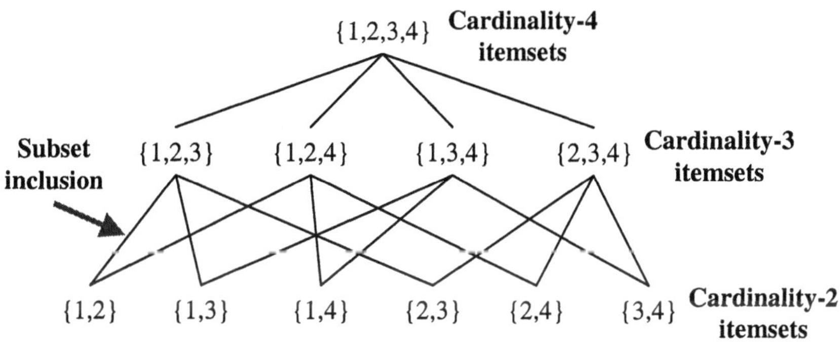

Figure 9: Itemset lattice for a set of 4 documents.

Itemset cardinality corresponds to a single level of the Hasse diagram. For itemset cardinality $|I|$, the number of possible itemsets is

$$\binom{n}{|I|} = \frac{n!}{|I|!(n-|I|)!}, \tag{4}$$

for n documents. The total number of possible itemsets over all cardinalities is 2^n.

In our matrix formalism, itemset supports are computed for sets of columns (cited documents) of the adjacency matrix, just as they are computed for pairs of columns in computing co-citation counts. For itemset I of cardinality $|I|$, whose member documents correspond to columns $j_1, j_2, \cdots, j_{|I|}$, its scalar support $\zeta(I)$ is

$$\zeta(I) = \sum_i a_{i,j_1} a_{i,j_2} \cdots a_{i,j_{|I|}} = \sum_i \prod_{\alpha=1}^{|I|} a_{i,j_\alpha}, \tag{5}$$

where i indexes rows (citing documents). Just as for pairwise co-citations, the term $a_{i,j_1}a_{i,j_2}\cdots a_{i,j_{|I|}}$ represents single co-citation occurrences, which are now generalized to higher orders. The summation then counts the individual co-citation occurrences.

A central problem in data mining is the discovery of frequent itemsets, i.e. itemsets with support greater than some threshold amount. In the context of linked documents, such frequent itemsets represent groups of highly similar documents based on higher- order co-citations. But managing and interacting with itemsets for document retrieval is problematic. Because of the combinatorially exploding numbers of itemsets and their overlap, user interaction becomes unwieldy.

Also, standard tools of analysis and visualization such as clustering and the minimum spanning tree assume an input matrix of pairwise distances. Mathematically, distances for all document pairs correspond to a fully connected distance graph. But the generalization to higher-order distances means that the distance graph edges are generalized to hyperedges, that is, edges that are incident upon more than two vertices. It is difficult to generalize clustering or minimum spanning tree algorithms to such distance hypergraphs.

We solve this dilemma by applying standard clustering or minimum spanning tree algorithms, but with pairwise distances that include higher-order co-citation similarities. The new distances we propose are thus a hybrid between standard pairwise distances and higher-order distances. For information retrieval visualization, users need only deal with disjoint sets of items, rather than combinatorial explosions of non-disjoint itemsets. The approach is designed such that member documents of frequent itemsets are more likely to appear together in clusters.

The following is the chain of reasoning leading to these link-mining distances. Consider the lattice of itemsets partially ordered by the set inclusion relation. Associated with each itemset in the lattice is its support. Consider any cardinality-2 itemset (document pair) $\{j,k\}$. Itemsets I that are proper supersets of $\{j,k\}$ are greater than $\{j,k\}$ in terms of the partial ordering, that is, $\{j,k\} \subset I \Leftrightarrow \{j,k\} < I$.

All the information at our disposal about the pairwise distance $d(j,k) = d(k,j)$ is then contained in the supports for these itemsets: $\{j\}$, $\{k\}$, $\{j,k\}$, and all itemsets $I > \{j,k\}$. The traditional way to compute the distance $d(j,k)$ is based simply on the support of $\{j,k\}$ as a measure of similarity. After normalization, the support is converted from a similarity to a dissimilarity (distance) via multiplicative or additive inversion.

A straightforward model for higher-order document similarities is to simply sum supports over all itemsets that contain the document pair in question. The itemset supports can thus be interpreted as features. More formally, the itemset support feature summation is

$$s_{j,k} = \sum_{\{I:j,k \epsilon I\}} \zeta(I). \tag{6}$$

This yields the similarity $s_{j,k}$ between documents j and k, where $\zeta(I)$ is the support of itemset I.

It is not feasible to compute all possible itemset supports. In practice, a reasonable approximation is to compute only those itemsets above a certain minimum support $\zeta_{\min}$. The document similarity in (6) then becomes

$$s_{j,k} = \sum_{\{I:j,k \epsilon I; \zeta(I) \geq \zeta_{\min}\}} \zeta(I). \tag{7}$$

By doing this, we can take advantage of fast algorithms (e.g. the *Apriori* algorithm [2]) for computing so-called frequent itemsets.

Algorithms for computing frequent itemsets are generally based on two principles. The first is that every subset of a frequent itemset is also frequent, so that higher- cardinality itemsets need only be considered if all their subsets are frequent. The second is that given a partitioning of the set of citing documents, an itemset can be frequent only if it is frequent in at least one partition, allowing the application of divide-and-conquer algorithms.

The worst-case complexity of the frequent-itemset problem is exponential. But empirically, fast algorithms scale linearly with respect to both the number of citing documents and the number of documents each of these cite. In fact, empirically these algorithms actually speed up slightly as the total size of the cited document collection increases.

If (7) is applied directly as a clustering similarity, experience has shown that the resulting clusters match poorly with frequent itemsets. In other words, this similarity formula is poor at preserving the higher-order co-citations (citation frequent itemsets). But an extension of this formula has been proposed [11] that yields clusters that are generally consistent with frequent itemsets.

In particular, a nonlinear transformation $T[\zeta(I)]$ is applied to the itemset supports $\zeta(I)$ before summation. The transformation T must be super-linear (asymptotically increasing more quickly than linearly), so as to favor large itemset supports. Example transformations include raising to a power greater than unity, or an increasing exponential. The document similarity in (7) then becomes

$$s_{j,k} = \sum_{\{I:j,k\epsilon I;\zeta(I)\geq\zeta_{\min}\}} T[\zeta(I)]. \tag{8}$$

The similarity $s_{j,k}$ is then normalized via

$$\hat{s}_{j,k} = \frac{s_{j,k} - \min(s_{j,k})}{\max(s_{j,k}) - \min(s_{j,k})}, \tag{9}$$

yielding the normalized similarity $\hat{s}_{j,k}\epsilon[0,1]$. Finally, this is converted from similarity to distance via additive inversion:

$$d_{j,k} = 1 - \hat{s}_{j,k}. \tag{10}$$

The resulting distance formula has the properties we desire. In particular, it includes higher-order co-citation information (association mining itemset supports), while retaining a simple pairwise structure. This provides a much stronger link-based clustering criterion, while enabling the direction application of standard similarity-based clustering algorithms.

3.2 Example Application of Link-Mining Distances

Let us examine an example application of the document distances in (10). We take this example from a set of documents retrieved from the Science Citation Index [18], which contains citation links among all documents in its collection. The query for this example is keyword "microtubules" for year 1999.

We select the first 100 documents returned from the query, which together cite 3070 unique documents. We retain only those 45 documents that have been cited 6 or more times, to retain only the more frequently cited documents. Figure 10 shows the resulting 100 × 45 citation matrix. In the figure, black indicates presence of citation and white indicates lack of citation.

Figure 11 shows a distribution of itemset supports for the "microtubules" document set. In particular, the left-hand side shows the support for each 3-itemset (itemset of cardinality 3), and the right-hand side shows the log-histogram of those supports. Cardinality-3 itemsets were chosen as representative of the general form that itemset distributions take. In fact, this support distribution is representative not only of itemsets of a particular cardinality, but for general document sets as a whole.

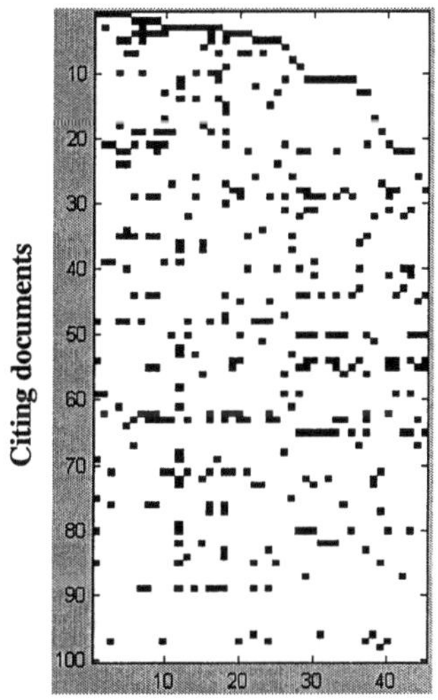

Figure 10: Citation matrix for SCI "microtubules" document set.

Figure 11 also includes the positive side of the Laplacian distribution, which appears linear on the log scale. The general Laplacian distribution $p(x)$ is

$$p(x) = \frac{\lambda}{2} e^{-\lambda |x|}. \tag{11}$$

The distribution is parameterized by $\lambda = \sqrt(2)/\sigma$, for standard deviation σ. Since itemset support is always positive, only the positive side of the distribution is needed, i.e.

$$p^{+}(x) = \lambda e^{-\lambda x}, x \geq 0. \tag{12}$$

As in Figure 11, Figure 12 shows cardinality-3 itemset supports for the "microtubules" document set. But this time, the itemset supports are non-linearly transformed, and only frequent itemsets (those with support above a particular threshold) are included. This shows the actual itemset support values that appear in the document similarity computation of (8).

It is important to point out the effect that the Laplacian itemset support distribution has on the computational complexity of our approach. Consider that fast algorithms for frequent itemsets compute supports starting from

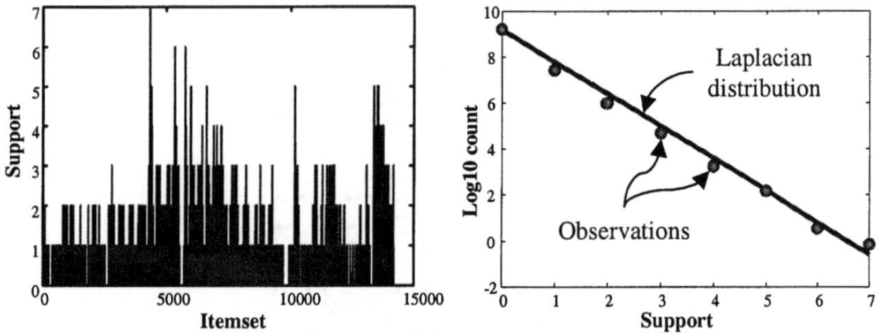

Figure 11: Itemset-support distribution for "microtubules" document set.

lower cardinalities. In going to the next higher cardinality, they consider only supersets of those itemsets found to be frequent at the lower cardinality. Because of the very heavy concentration of smaller supports in the Laplacian distribution, a large portion of itemsets are below the minimum support threshold, and get removed from any further consideration at higher cardinalities. This in turn contributes to low average computational complexity.

The Laplacian distribution of itemset supports is also consistent with the fact that our approach yields clusters that tend to be consistent with more frequent itemsets. Frequent itemsets are already sparse for this distribution, and the nonlinear transformation in (8) increases the sparseness. In fact, we can give a theoretical guarantee for the consistency between clusters and frequent itemsets. This guarantee holds independent of any particular clustering criterion.

Details of the proof are in [10]. The key idea of the proof is that for distances computed with (8) through (10), the most frequent itemset can always be made a cluster, given a large enough degree of nonlinearity in the transformation of itemset supports. This relies on the fact that for a super-linear transformation $T = T_p = T(\zeta; p)$ of itemset supports ζ in (8), as the degree of nonlinearity p increases, $T_p(\zeta)$ with a larger ζ is asymptotically bounded from below by $T_p(\zeta)$ with a smaller ζ. Since the term with largest ζ asymptotically dominates the distance summation, the result is that documents in the most frequent itemset are asymptotically closer to one another than to any other documents, thus forming a cluster.

We then generalize the proof to cover the clustering of arbitrary itemsets

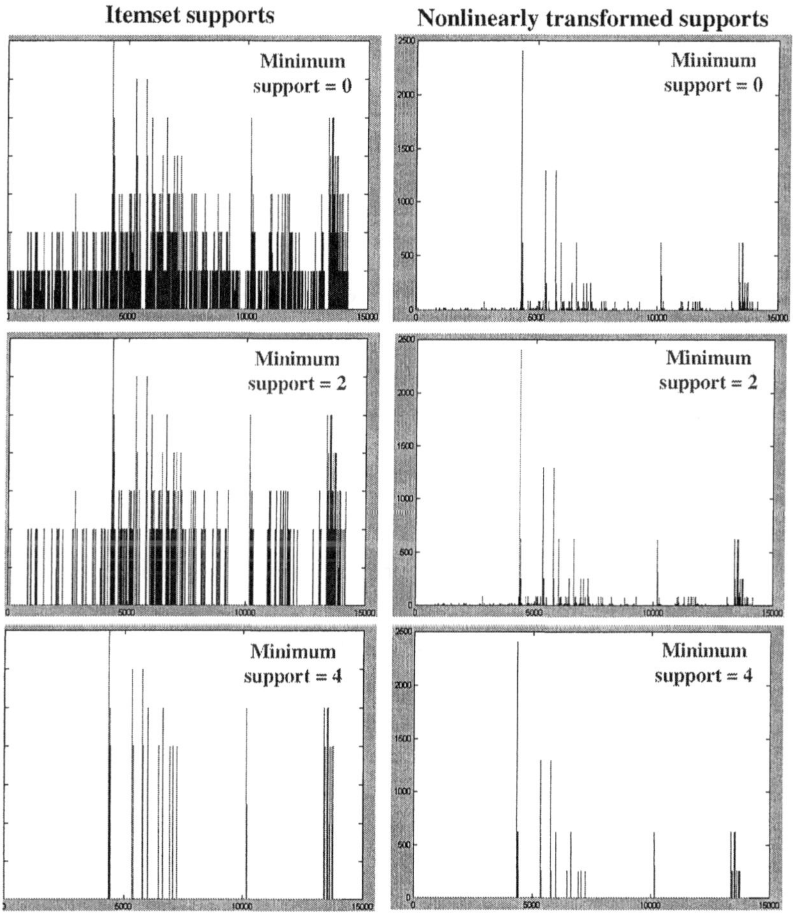

Figure 12: Transformed frequent itemsets for "microtubules" document set.

in terms of their relative supports and document overlap. The result is that more frequent itemsets asymptotically form clusters at the expense of less frequent itemsets that overlap them. If there is no overlapping itemset with more support, then a given itemset will form a cluster for a sufficiently large value of the nonlinearity parameter p. For overlapping itemsets with equal support, the itemset members not in common asymptotically form clusters, and each of the members in common are asymptotically in one of those clusters, though there is no guarantee for exactly which of them.

Our theoretical result provides no such guarantee for itemsets of equal support. More importantly, it provides no upper bound on the necessary value of p to ensure itemset clustering for a given data set. But empirically, we have found [11] that the transformation

$$T(\zeta) = \zeta^p \tag{13}$$

with $p = 4$ usually results in the most frequent itemsets appearing together in clusters.

4 Link Mining for Hierarchical Document Clustering

For document retrieval, results from simple keyword queries can be clustered into more refined topics. Such clustering can help users identify subsets of the retrieval results that are more likely to contain relevant documents. Indeed, studies have shown that precision and recall tends to be higher within the best cluster than within the retrieval results as a whole [19]. Users can thus concentrate on clusters with higher proportions of relevant documents, avoiding clusters with mostly non-relevant documents. Users can search for better clusters by looking at small samples from each cluster.

Additionally, document clustering can be applied hierarchically, yielding clusters within clusters. This facilitates a large-to-small-scale analysis of retrieval results reminiscent of binary search. This is based on the idea that larger scale clusters correspond to more general topics, and smaller scale clusters correspond to more specific topics within the general topics. Cluster hierarchies thus serve as topic hierarchies. For information retrieval, content analysis of document samples within large-scale clusters allows the user to choose among general topics. Once a particular general topic is chosen, the user can sample documents within smaller-scale clusters within the large- scale cluster. The user can then proceed with this analysis recursively, progressively refining the topic.

4.1 Hierarchical Clustering, Dendrograms, and Document Retrieval

The dendrogram is a tree visualization of a hierarchical clustering. Leaves of the dendrogram tree are individual documents, at the lowest level of the hierarchy. Non-leaf nodes represent the merging of two or more clusters, at increasing levels of the hierarchy. A node is drawn as a horizontal line that spans over its children, with the line drawn at the vertical position corresponding to the merge threshold distance. The dendrogram visualization is illustrated in Figure 13.

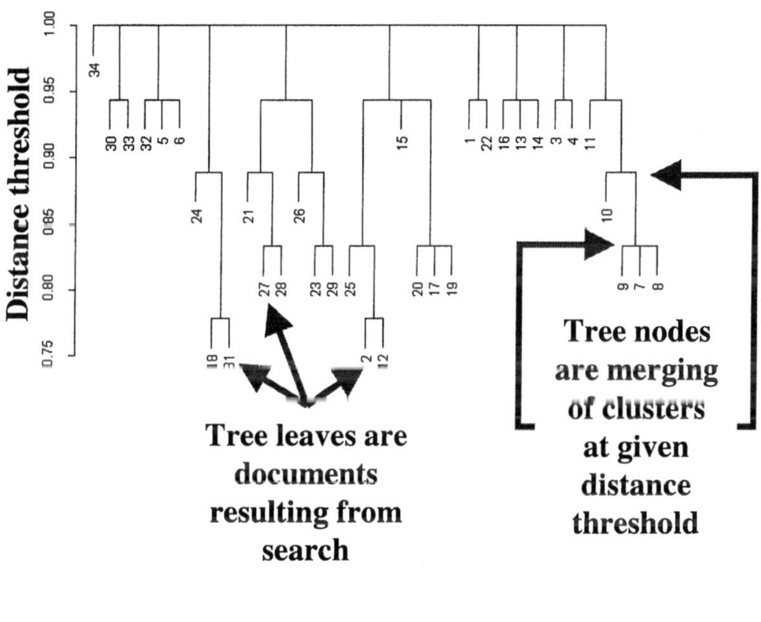

Figure 13: Dendrogram for visualizing hierarchical document clusters.

For a given hierarchical clustering, the clusters resulting from a given threshold distance can be readily determined from the dendrogram (see Figure 14). If a horizontal line is envisioned at the given threshold value, tree nodes directly below the threshold define clusters. That is, the nodes' children are all members of the same cluster. Nodes that lie above the threshold each represent clusters of single documents.

The dendrogram serves as the basis for a graphical user interface for information retrieval. The hierarchical display facilitates interactive large-

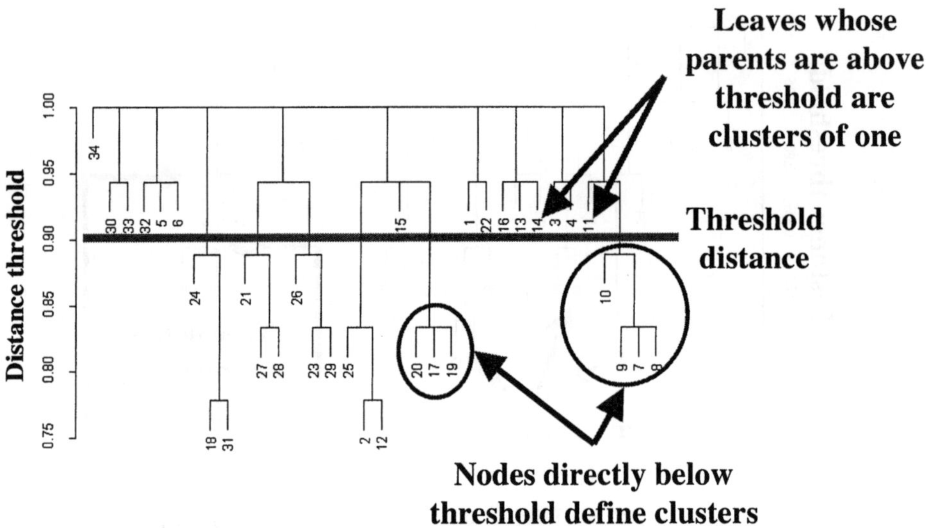

Figure 14: Interpreting clusters from dendrogram for given clustering threshold distance.

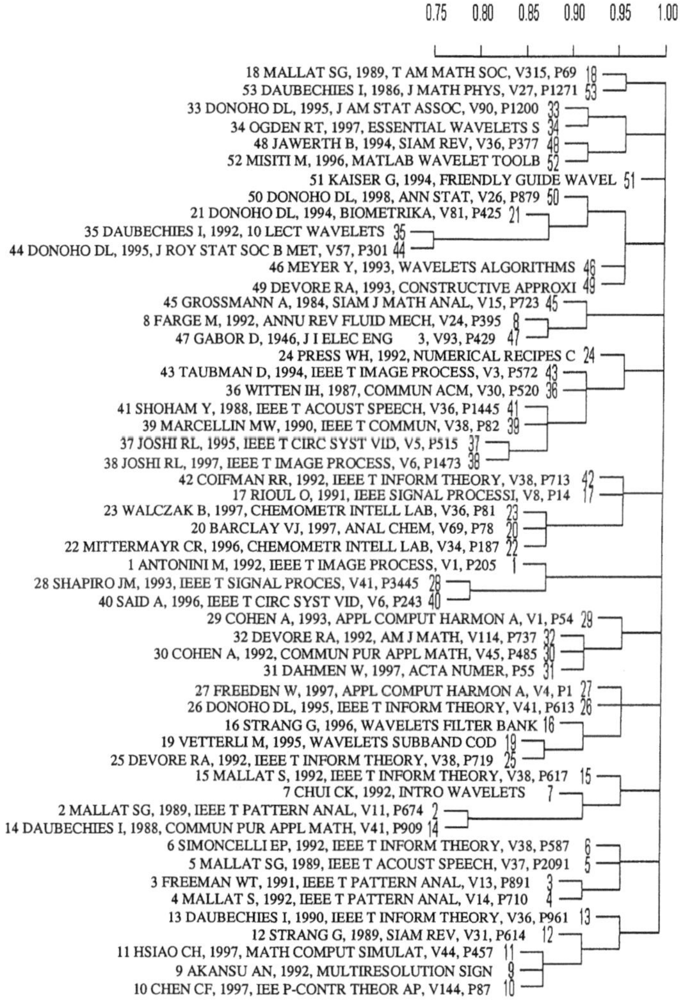

Figure 15: Dendrogram rotated and augmented with text for document retrieval.

to-small-scale cluster analysis for topic refinement. By selecting a leaf node, the user can retrieve metadata for an individual document, or can retrieve the document itself.

For a document retrieval user interface, the dendrogram can be rotated by 90 degrees and augmenting with text, as shown in Figure 15. In the usual dendrogram orientation, there is an arbitrary amount of blank space available below each leaf, in the vertical direction. The 90-degree rotation provides arbitrary space available in the horizontal direction, consistent with lines of text. Within information retrieval, the general approach of clustering for topic refinement is well established. But for document searching, some questions have been raised about the value added by clustering visualizations in comparison to the traditional query approach [19].

Clustering visualizations have obvious value in providing thematic overviews of document collections, particularly useful for browsing. But for the search process, the lack of textual representations such as titles and topical words can make systems difficult to use. In general, it is difficult to understand document contents without actually reading some text. The situation is compounded by the fact that it is difficult to objectively measure visualization effectiveness [7].

4.2 Illustrative Example

We now demonstrate our approach to link-based document clustering for an example document set. In particular, we compare link association mining results to clusters resulting from our link-based similarity formula (8). The example document set is from the Science Citation Index (SCI) [18].

We do the itemset/clustering comparison by a novel augmentation of the dendrogram with members of frequent itemsets. This allows an easy visual assessment of itemset/cluster consistency. For this example, we do an SCI query with keyword "wavelet*" for the year 1999. The first 100 documents returned by the query cite 1755 documents. We filter these cited documents by citation count, retaining only those cited three or more times, resulting in a set of 34 highly cited documents. We then compute complete-linkage, average-linkage, and single-linkage clusters for the set of 34 highly cited documents. Here we first apply the traditional pairwise method of computing co-citation based distances. The resulting augmented dendrogram is shown in Figure 16. The dendrogram is augmented by the addition of graphical symbols for members of frequent 4-itemsets, added at the corresponding tree leaves. In this example, the most frequently occurring 4-itemset is {2, 17, 19, 20} . For complete linkage, documents 17, 19, and 20 of this itemset are a possible cluster. These documents apply wavelets to problems in the field of chemistry, and are well separated from the rest of the collection, both thematically and in terms of co-citations. But including document 2 (a foundational wavelet paper by wavelet pioneer Mallat) in this cluster would require the inclusion of documents 12, 15, and 25, which are not in the itemset. These three additional documents are another foundational wavelet paper by Mallat, and two foundational papers by fellow pioneer Daubechies.

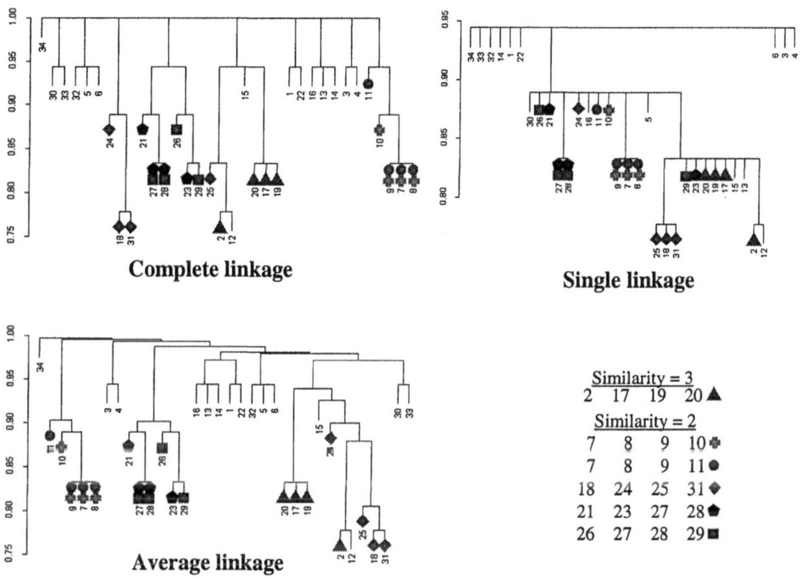

Figure 16: Inconsistency between clusters and frequent itemsets for pairwise document distances.

For single linkage, there is even less cluster/itemset consistency. The itemset {2, 17, 19, 20} is possible within a cluster only by including 8 other documents. We interpret this as being largely caused by single linkage chaining. In general, the application of clustering to mere pairwise co-citation similarities is insufficient for ensuring that itemsets of larger cardinality appear as clusters, even with complete-linkage. The overlap for the 4-itemsets {7, 8, 9, 10} and {7, 8, 9, 11} , corresponds to the 5-itemset {7, 8, 9, 10, 11}. Thematically, these 5 papers are largely foundational. The combined two 4-itemsets are a complete-linkage cluster. But for single-linkage, 24 other documents would need to be included in order for the two itemsets to be a cluster. Again, pairwise clustering is a necessary but insufficient condition for frequent itemsets. We have a similar situation for the 4-itemsets {21, 23, 27, 28} and {26, 27, 28, 29} , though with a lesser degree of itemset overlap. Thematically, these papers are applications of wavelets in image coding.

For the 4-itemset {18, 24, 25, 31} , three of the papers are by Donoho, who works in wavelet-based statistical signal estimation for denoising. These three papers are a complete-linkage cluster, as well as a single-linkage cluster. The remaining document in the 4-itemset is a foundational book by Daubechies. Including it in a complete-linkage cluster would require the inclusion of every document in the set, while including it in a single-linkage cluster would require the inclusion of 21 other documents.

As a comparison with traditional pairwise co-citation clustering, Figure 17 shows clusters for our distances that include link-mining itemset supports. In particular, it shows complete-linkage clusters with document distances computed via (8) through (10). There are three separate cases, each case being taken over multiple values of itemset cardinality χ. The three cases are $\chi = 2, 3$; $\chi = 2, 3, 4$; $\chi = 3, 4$. Here the itemset supports $\zeta(I)$ are nonlinearly transformed by $T[\zeta(I)] = [\zeta(I)]^4$.

Consistency between clusters and frequent itemsets is nearly perfect with our link-mining distances. The most frequent itemset {2,17,19,20} forms a cluster for two of the cases ($\chi = 2, 3, 4$ and $\chi = 3, 4$). The source of the inconsistency for the case $\chi = 2, 3$ is apparently the lowest-order (pairwise) supports. Lower order supports are generally larger than higher-order supports, and thus tend to dominate the summation in (8). All other frequent itemsets are consistent with these clusters, at least to the extent possible given their overlap. That is, {7, 8, 9, 10} overlaps with {7, 8, 9, 11} and {21, 23, 27, 28} overlaps with {26, 27, 28, 29}, which prevents them from forming individual clusters.

4.3 Itemset-Matching Clustering Metric

In comparing clustering to association mining itemsets, the important issue is whether frequent itemsets form clusters comprised only of the itemset members. This is equivalent to determining the minimal-cardinality cluster that contains all the members of a given itemset and then comparing that cluster cardinality to the itemset cardinality. The portion of a minimal cluster occupied by an itemset could serve as an itemset-matching metric for a clustering. Moreover, it could be averaged over a number of itemsets to yield an overall itemset-matching metric for a clustering.

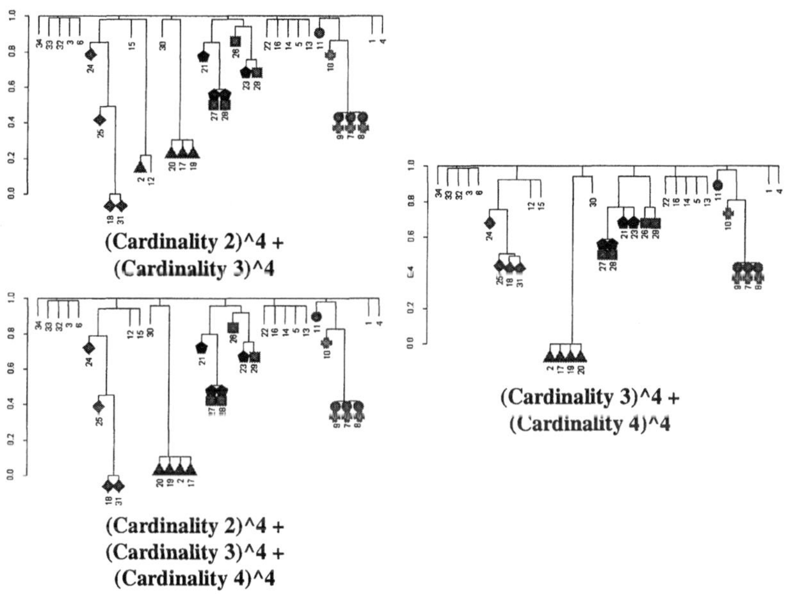

Figure 17: Clusters from our link-mining distances are much more consistent with frequent itemsets.

We describe this itemset-matching metric more formally. Let

$$\pi = \{\pi_1, \pi_2, \cdots, \pi_{k_1}\}$$

be a partition of items (documents) that is consistent with a hierarchical clustering merge tree. Furthermore, let $I = \{I_1, I_2, \cdots, I_{k_2}\}$ be a set of itemsets. Then for each itemset $I_i \epsilon I$, there is some block of the partition $\pi_j \epsilon \pi$ such that $|\pi_j|$ is minimized, subject to the constraint that $I_i \subseteq \pi_j$. We call this π_j the minimal cluster containing the itemset. The fact that such a minimal cluster exists can be proven by straightforward induction. The constraint $I_i \subseteq \pi_j$ is satisfied trivially for a partitioning in which a single block contains all items in the original set, corresponding to the highest level of the merge tree.

Moving down to the next highest level of the merge tree, either some block of the partition $\pi_j \epsilon \pi$ satisfies $I_i \subseteq \pi_j$, or else not. If not, then the block in the highest-level partition is the minimal cluster containing the itemset. Otherwise this process can be repeated, until a level is reached in which the constraint $I_i \subseteq \pi_j$ fails. At this point, the minimal cluster containing the itemset is found from the previous level, as the one in which $I_i \subseteq \pi_j$. A similar argument can start from the leaves of the merge tree and proceed upward.

Once a minimal (cardinality) cluster π_j is found for an itemset, a metric can be defined for measuring the extent to which the itemset is consistent with the cluster. This metric $M(\pi, I_i)$ is simply the portion of the cluster occupied by the itemset, or in terms of set cardinalities,

$$M(\pi, I_i) = \frac{|I_i|}{|\pi_j|}.$$

Again, this requires that $|\pi_j|$ be minimized, for $\pi_j \epsilon \pi$, subject to the constraint $I_i \subseteq \pi_j$, and π is consistent with the merge tree. The metric $M(\pi, I_i)$ is defined for a set of itemsets I by averaging $M(\pi, I_i)$ over $I_i \epsilon I$, that is,

$$M(\pi, I) = \frac{1}{|I|} \sum_{I_i \epsilon I} M(\pi, I_i) = \frac{1}{|I|} \sum_{I_i \epsilon I} (\frac{|I_i|}{|\pi_j|}). \tag{14}$$

The itemset-matching metric $M(\pi, I)$ takes its maximum value of unity when $I_i = \pi_j$, indicating the best possible match between itemsets and clusters. The proof is that since $|I_i| = |\pi_j|$,

$$M(\pi, I) = \frac{1}{|I|} \sum_{I_i \epsilon I} 1 = \frac{|I|}{|I|} = 1.$$

The minimum value of $M(\pi, I)$ is $M(\pi, I) = |I_i|/n$, indicating the poorest possible match. For the proof, consider that $M(\pi, I) = |I_i|/|\pi_j|$ for a given $|I_i|$ takes its minimum value when $|\pi_j|$ takes its maximum value of $|\pi_j| = n$. Then the minimum $M(\pi, I)$ is the sum of minimum $M(\pi, I_i)$, that is,

$$M(\pi, I) = \frac{1}{|I|}\sum_{I_i \epsilon I}(\frac{|I_i|}{|\pi_j|}) = \frac{1}{|I|}\sum_{I_i \epsilon I}(\frac{|I_i|}{n}) = \frac{1}{|I|}\frac{|I_i|}{n}\sum_{I_i \epsilon I} 1 = \frac{|I|}{|I|}\frac{|I_i|}{n} = \frac{|I_i|}{n}.$$

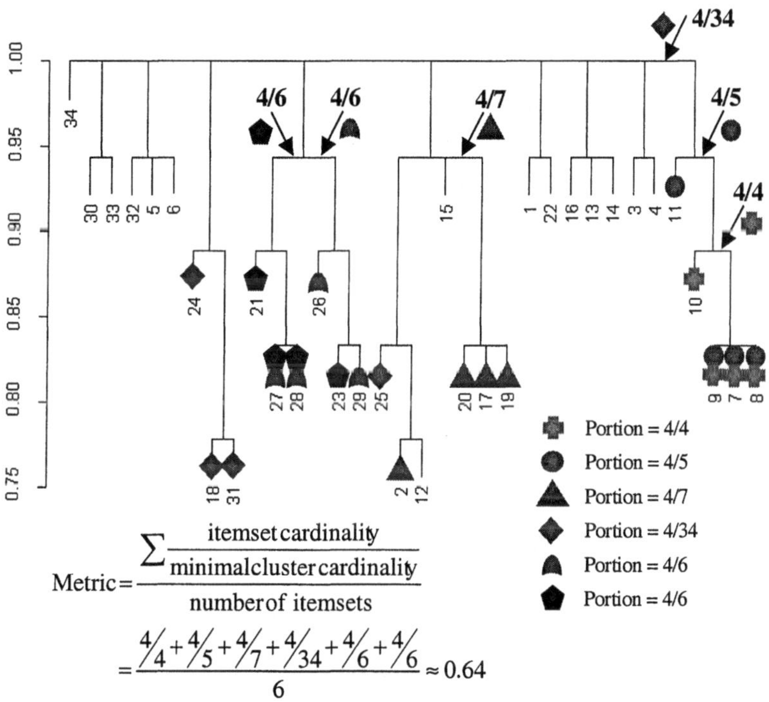

Figure 18: Itemset-matching clustering metric.

Figure 18 illustrates the itemset-matching clustering metric $M(\pi, I)$. For a given itemset, there is some minimal threshold value at which it is a subset (not necessarily proper) of a cluster. For this threshold value, the cluster may contain documents other than those in the itemset (in which case the itemset is a proper subset of the cluster). The ratio of the itemset cardinality to cluster cardinality is the size of the itemset relative to the cluster size.

The metric is then the average of these relative itemset sizes over a set of itemsets.

4.4 Experimental Validation

In this section, we apply our proposed document distances to real-world document citations. In particular, we apply them to data sets from the Science Citation Index (SCI). Science citations are a classical form of hypertext, and are of significant interest in information science. The assumption that citations imply some form of topical influence holds reasonably well. That is, for citations all links between documents have the same general semantics. This is in contrast to Web documents, in which links could be for a variety of other purposes, such as navigation. The general method of clustering documents and computing itemset-matching clustering metrics is shown in Figure 19. We include higher-order co-citations in pairwise document distances, and compute standard co-citation distances as a comparison. We also compute frequent itemsets of given cardinalities and minimum supports. Once a clustering is computed for the given distance function, we subject it to the itemset-matching metric for the given itemsets.

The SCI data sets we employ are described in Table 1. For each data set, the table gives the query keyword and publication year(s), the number of citing documents resulting from the query, and the number of documents they cite after filtering by citation count. For the data sets 1, 5, and 9, results are included for both co-citations and bibliographic coupling, yielding data sets 2, 6, and 10 (respectively), for total of 10 data sets.

Our empirical tests apply the metric proposed in Section 4.3. The metric compares clustering to frequent itemsets, determining whether given itemsets form clusters comprised only of the itemset members. In other words, the metric determines the minimal-cardinality cluster that contains all the members of a given itemset, and compares that cluster cardinality to the itemset cardinality. This is then averaged over a number of itemsets, to yield an overall itemset-matching metric for a clustering.

Table 2 shows how link-mining distances are computed for the experiments with SCI data sets. The table shows the itemset cardinalities $|I|$ that are applied in the similarity formula (2), and the values of itemset support nonlinearity parameter p for itemset nonlinearity $T(\zeta) = \zeta^p$. For each link-mining distance formula, we compare metric values to those for standard pairwise distances. Here we apply the same cardinalities in the metric formula (14) as in the distance formula (8), to enable a direct interpretation of the results.

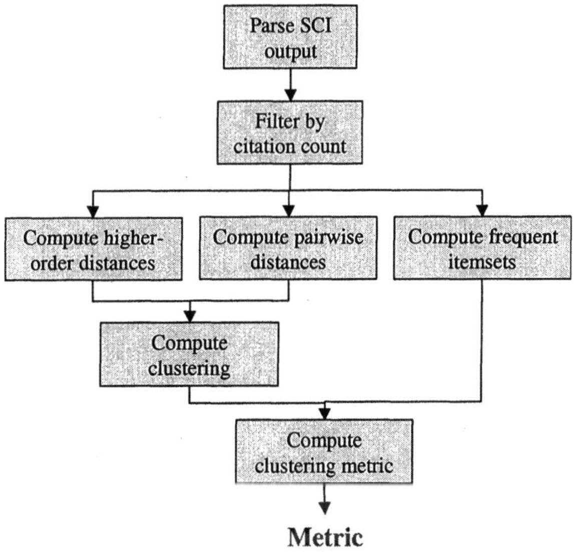

Figure 19: General method for computing itemset-matching clustering metric.

Table 1: Details for Science Citation Index data sets.

Data Set(s)	Query Keyword(s)	Year(s)	Citing Docs	Cited Docs
1, 2	adaptive optics	2000	89	60
3	collagen	1975	494	53
4	genetic algorithm* AND neural network*	2000	136	57
5,6	quantum gravity AND string*	1999-2000	114	50
7	wavelet*	1999	100	34
8	wavelet*	1999	472	54
9,10	wavelet* AND brownian	1973-2000	99	59

Table 2: Itemset cardinalities and support nonlinearities for link- mining distances.

Data Set(s)	[Itemset Cardinality, Support Nonlinearity]
1	[3,4], [3,6], [4,4], [4,6]
2,6	[3,4], [4,4], [4,6]
3,5,7,8,9,10	[3,4], [4,4]
4	[3,4], [3,6], [4,4]

The comparisons are done for each combination of complete-linkage, average- linkage, and single-linkage clustering. We also compare for each combination of the most frequent itemset, the 5 most frequent itemsets, and the 10 most frequent itemsets in the metric formula (14). For test cases in which nonlinearity parameter value $p = 4$ yields relatively low metric values, we include additional test cases with $p = 6$. Since there are 25 different combinations of itemset cardinalities and support nonlinearities, each with $3 \times 3 = 9$ different combinations of clustering criteria and sets of most frequent itemsets, there are $9 \times 25 = 225$ total test cases.

Table 3 compares clustering metric values for the test cases in Table 2. The table classifies cases as having metric values for link-mining distances equal to, greater than, or less than metric values for standard pairwise distances. Higher metric values correspond to clusters being more consistent with frequent itemsets. For most test cases (169 of 225 versus 27 of 225), the link-mining distances result in better consistency with frequent itemsets in comparison to standard pairwise distances. For a relatively small number of cases (29 of 225), the metric values are equal.

For the 20 combinations of 10 data sets and 2 itemset cardinalities, we find that itemset support nonlinearities with $p = 4$ are usually sufficient (15 of 20) for a good match to frequent itemsets. Otherwise nonlinearities with $p = 6$ in (13) are sufficient, in all but one of the 20 combinations. Here we consider a clustering metric value greater than about 0.7 to be a good match. This corresponds to a frequent itemset comprising on average about 70% of a cluster that contains all its members.

For data sets 3, 5, 8, 9, and 10, we compute itemset-matching clustering metrics resulting from the complexity reduction technique in (7), and compare metric values to the full-complexity method in (6). The clustering metric results of the 90 cases from the five extracted data sets are summarized in Table 4 and Table 5. The results show that excluding itemset

Table 3: Clustering metric comparisons for standard pairwise (P.W.) vs. link-mining higher-order (H.O.) distances.

Data set	H.O. = P.W.	H.O. > P.W.	H.O. < P.W.	Cases
1	6	16	14	36
2	7	15	5	27
3	0	18	0	18
4	1	24	2	27
5	3	13	2	18
6	2	22	3	27
7	2	16	0	18
8	5	13	0	18
9	3	14	1	18
10	0	18	0	18
Total	29	169	27	225

supports below itemset minimum support *minsup* generally has little effect on clustering results, particular for smaller values of *minsup*. However, there is some degradation in metric values for higher levels of *minsup*. Here "degradation" means that metric values are smaller when some itemset supports are excluded, corresponding to a poorer clustering match to frequent itemsets.

We offer the following interpretation for the *minsup*-dependent degradation in clustering metric. Members of frequent itemsets are typically frequently cited documents overall. Such frequently cited documents are likely to appear in many itemsets, even less frequent itemsets. Thus there are likely to be many itemsets below *minsup* that contain these frequently cited documents. Excluding itemsets below *minsup* then removes the supports that these itemsets contribute to the summations in computing link-mining distances.

5 Conclusions

In this chapter, we described new methods for enhanced understanding of relationships among documents that are returned by information retrieval systems based on a link or citation analysis. A central component in the methodology is a new class of inter-document distances that includes in-

Table 4: Clustering metrics for link-mining distances with full computational complexity (minsup 0) and reduced complexity (minsup 2).

Data set	(minsup 2) = (minsup 0)	(minsup 2) > (minsup 0)	(minsup 2) < (minsup 0)	Cases
3	18	0	0	18
5	18	0	0	18
8	18	0	0	18
9	18	0	0	18
10	11	0	7	18
Total	83	0	7	90

Table 5: Clustering metrics for link-mining distances with full computational complexity (minsup 0) and reduced complexity (minsup 4).

Data set	(minsup 4) = (minsup 0)	(minsup 4) > (minsup 0)	(minsup 4) < (minsup 0)	Cases
3	12	2	4	18
5	11	1	6	18
8	10	0	8	18
9	18	0	0	18
10	12	2	4	18
Total	63	5	22	90

formation mined from a hypertext collection. These distances rely on a higher-order counterparts of the familiar co-citation similarity, in which co-citation is generalized from a relationship between a pair of documents to one between arbitrary numbers of documents. These document sets of larger cardinality are equivalent to itemsets in association mining. Our experimental results show that in comparison to standard pairwise distances, our higher-order distances are much more consistent with frequent itemsets.

We also presented the application of the hierarchical clustering dendrogram for information retrieval. The dendrogram enables quick comprehension of complex query-independent relationships among the documents, as opposed to the simple query-ranked lists usually employed for presenting search results. We introduced new augmentations of the dendrogram to support the information retrieval process, by adding document-descriptive text and glyphs for members of frequent itemsets.

This work represents the original application of association mining in finding frequent itemsets for the purpose of visualizing hyperlink structures in information retrieval search results. The generalization of co-citation to higher orders helps prevent the obscuring of important frequent itemsets that often occurs with traditional co-citation based analysis, allowing the visualization of collections of frequent itemsets of arbitrary cardinalities. This work also represents a first step towards the unification of clustering and association mining.

References

[1] R. Agrawal, T. Imilienski, and A. Swami, "Mining Association Rules between Sets of Items in Large Databases," in *Proc. of the ACM SIGMOD Int'l Conf. on Management of Data*, May 1993, pp. 207-216.

[2] R. Agrawal, R. Srikant, "Fast Algorithms for Mining Association Rules," in *Proceedings of the 20th International Conference on Very Large Databases*, Santiago, Chile, September 1994, pp. 487-499.

[3] C. Chen, L. Carr, "Trailblazing the Literature of Hypertext: An author co- citation analysis (1989-1998)," in *Proceedings of the 10th ACM Conference on Hypertext (Hypertext '99)*, Darmstadt, Germany, February, 1999, pp. 51- 60.

[4] I. Daubechies, *Ten Lectures on Wavelets*, SIAM, Philadelphia, 1992.

[5] T. Fruchterman, E. Reingold, "Graph Drawing by Force-Directed Placement," *Software—Practice and Experience*, 21, pp. 1129-1164, 1991.

[6] E. Garfield, M. Malin, H. Small, "Citation Data as Science Indicators," in *Toward a Metric of Science: The Advent of Science Indicators*, Y. Elkana, J. Lederberg, R. Merten, A. Thackray, H. Zuckerman (eds.), John Wiley & Sons, New York, 1978, pp. 179-207.

[7] I. Herman, G. Melanon, M. Marshall, "Graph Visualization and Navigation in Information Visualization: a Survey," *IEEE Transactions on Visualization and Computer Graphics*, 6(1), pp. 24-43, 2000.

[8] J. Kleinberg, "Authoritative Sources in a Hyperlinked Environment," in *Proceedings of the ACMSIAM Symposium on Discrete Algorithms*, January 1998, pp. 668-677.

[9] S. Noel, H. Szu, "Multiple-Resolution Clustering for Recursive Divide and Conquer," in *Proceedings of Wavelet Applications IV*, Orlando, FL, April 1997, pp. 266-279.

[10] S. Noel, *Data Mining and Visualization of Reference Associations: Higher Order Citation Analysis*, Ph.D. Dissertation, Center for Advanced Computer Studies, The University of Louisiana at Lafayette, 2000.

[11] S. Noel, V. Raghavan, C.-H. H. Chu, "Visualizing Association Mining Results through Hierarchical Clusters," in *Proc. First IEEE International Conference on Data Mining*, San Jose, California, November 2001, pp. 425- 432.

[12] L. Page, S. Brin, R. Motwani, T. Winograd, *The PageRank Citation Ranking: Bringing Order to the Web*, Stanford Digital Library, Working Paper 1999- 0120, 1998.

[13] H. Small, "Co-citation in the Scientific Literature: A New Measure of the Relationship Between Two Documents," *Journal of the American Society of Information Science*, 24, pp. 265-269, 1973.

[14] H. Small, "Macro-Level Changes in the Structure of Co-Citation Clusters: 1983-1989," *Scientometrics*, 26, pp. 5-20, 1993.

[15] G. Strang, T. Nguyen, *Wavelets and Filter Banks*, Wellesley-Cambridge, Wellesley, Massachusetts, 1996.

[16] W. Venables, B. Ripley, *Modern Applied Statistics with S-Plus*, Springer- Verlag, 1994.

[17] J. Wise, J. Thomas, K. Pennock, D. Lantrip, M. Pottier, A. Schur, V. Crow, "Visualizing the Non-Visual: Spatial Analysis and Interaction with Information from Text Documents," in *Proceedings of Information Visualization '95 Symposium*, Atlanta, GA, 1995, pp. 51-58.

[18] The ISI Science Citation Index (SCI), available through the ISI Web of Science, http://www.isinet.com/isi/products/citation/sci/.

[19] R. Baeza-Yates, B. Ribeiro-Neto (eds.), *Modern Information Retrieval*, Addison Wesley Longman, 1999.

CLUSTERING AND INFORMATION RETRIEVAL (pp. 195-225)
W. Wu, H. Xiong and S. Shekhar(Eds.)

Query Clustering in the Web Context

Ji-Rong Wen
Microsoft Research Asia
3F, Beijing Sigma Center
No. 49, Zhichun Road, Haidian District
Beijing, P.R. China 100080
E-mail: `jrwen@microsoft.com`

Hong-Jiang Zhang
Microsoft Research Asia
3F, Beijing Sigma Center
No. 49, Zhichun Road, Haidian District
Beijing, P.R. China 100080
E-mail: `hjzhang@microsoft.com`

Contents

1 Introduction

Query clustering is a class of techniques aiming at grouping users' semantically related, not syntactically related, queries in a query repository, which were accumulated with the interactions between users and the system. While there are numerous previous works on document clustering, query clustering is a relatively new topic. The driving force of the development of query clustering techniques comes recently from the requirements of modern web searching. Below we briefly analyze several motivations and applications of query clustering — FAQ detecting, index-term selection and query reformulation.

FAQ Detecting Faced with the increasing requirement for more precise information retrieval systems, a new generation of search engines — or question answering systems — has emerged on the Web (e.g. AskJeeves, http://www.askjeeves.com). Unlike traditional search engines, these new systems try to "understand" user's questions in order to suggest similar questions that other people often ask and for which the system has the correct answers. In fact, the correct answers have been prepared or checked by human editors in most cases. It is then guaranteed that, if one of the suggested questions is truly similar to that of the users, the answers provided by the system will be relevant. The assumption behind such a system is that many people are interested in the same questions — the Frequently Asked Questions (FAQs). It is assumed that if the system can correctly understand or identify these questions, then a significant majority of users' questions will be answered precisely. The human editors have to detect which questions/queries are FAQs from users' query logs so that they can prepare/check their answers. However, this is a tedious task. What is needed is an automatic tool that helps the editors to identify FAQs, so that the editors only need to decide the answers to those FAQs. This is exactly a query clustering problem — to cluster similar queries/questions together in order to discover FAQs. However, the queries submitted by web users are very different, and they are not always well-formed questions. Queries with the same or similar search intentions may be represented with diverse words. Therefore, some measures are needed to estimate the similarity between queries, which is the core problem of query clustering.

Index-Term Selection Despite the fact that keywords are not always good descriptors of the contents, most existing search engines still solely rely on keywords contained in web pages to construct indices. This is one of the key factors that affect the precision of search engines. Inconsistency between term usages in queries and documents is a well-known problem in information retrieval, which was first observed by Furnas [9]. It is even worse when the query is very short as often the case on the web. Through web logs, [3] testified the "word mismatching" assumption in quantity. The test results show that there is indeed a large difference between the query terms and document terms, and the traditional way of directly extracting index terms from documents will not be effective when the user submits queries containing terms different from those in the documents. Query clustering is a promising technique to provide a solution to the word mismatching problem. If similar queries can be recognized and clustered together, the resulting query clusters will be very good sources for selecting additional index terms for documents. For example, if queries "atomic bomb", "Manhattan Project" "Hiroshima bomb" and "nuclear weapon" are put into a query cluster, this cluster, not the individual terms, can be used as a whole to index documents related to atomic bomb. In this way, any queries contained in the cluster can be linked to these documents.

Query Reformulation The web is not a well-organized information source where innumerous "authors" create their websites independently. Those authors "vocabularies" vary greatly. Moreover, most words in the natural language have inherent ambiguity. These reasons make it rather difficult for the web users to formulate queries with appropriate words. Many web search engines, such as AltaVista (http://www.altavista.com), Excite (www.excite.com), Lycos (www.lycos.com), etc., attempt to identify some of the users' intentions and suggest a list of alternate terms for the user to reformulate his/her query. This kind of term suggestion mechanism needs to track the users emerging topics of interest, such as events in the news or newly publicized web sites. Obviously, query clustering, which intends to find related queries from users' daily logs, could be one of the critical techniques to drive this term suggestion process.

For any clustering problem, researches have been concerned mainly with two aspects — similarity functions and algorithms for the clustering process. The key problem underlying query clustering is to determine an adequate similarity function so that truly similar queries can be grouped together using a clustering algorithm. Currently, there are mainly two categories of methods to calculate the similarity between queries: one is based on query content, and the other on query session.

The classic approach to information retrieval (IR) would suggest a similarity calculation between queries according to their contents (e.g., keywords) [26]. The content-based query clustering methods just follow the basic term-semantic hypothesis in IR: *If two queries contain the same or similar terms, they denote the same or similar information needs.* However, this hypothesis has some known drawbacks due to the limitations of keywords, just as in the case of keyword-based IR. In the case of queries, in particular, the keyword-based similarity calculation will be very inaccurate (with respect to semantic similarities) due to the short length of queries.

To overcome the short-query problem, researchers are looking for other evidences to judge the similarity between queries [28][29][1][12]. The most promising way is to extend the "query" concept to the "query session" concept. A query session is defined as the query and the subsequent activities the user performed. Usually, query sessions can be extracted from the logs of search engines. The basic assumption here is that the activities following a query are relevant to the query and represent, to some extent, the semantic features of the query. The query text and the activities in a query session as a whole can represent the search intention of the user more precisely. Moreover, the ambiguity of some query terms is eliminated in query sessions. For instance, if the user visited a few tour websites after submitted a query "Java", it is reasonable to deduce that the user was searching for information about "Java Island" not "Java programming language". Accordingly, "query clustering" can be extended to "query session clustering". The hypothesis underlying the session-based query clustering methods is formally stated as: *Two queries leading to the similar user activities are similar.*

User activities are comparable to user relevance feedback in a traditional IR environment, except that user activities denote implicit and not always valid relevance judgments.

Similarities based on query contents and query sessions represent two different points of view. The two criteria have their own advantages and shortcomings. In general, content-based criterion tend to cluster queries with the same or similar terms; but, due to the ambiguity of words, similar terms could be used to represent different query requirements and different terms may represent the same requirements. Session-based criterion tend to cluster queries related to the same or similar topics; but a session is usually noisy and contains more than one topic or "interest point", thus queries with different intentions may lead to similar activities. Since information needs may be partially captured by both query contents and user activities, some combined measures taking advantage of both criteria may be more appealing.

2 Content-based Query Clustering

Although the discussion of query clustering is relatively new, there have been extensive studies on document clustering. The first group of related clustering approaches is certainly those that cluster documents using the contents they contain. Like documents, there are different ways to represent query content: keywords, words in their order, and phrases. They provide different measures of similarity.

2.1 Similarity Based on Keywords or Phrases

The first measure of query similarity comes directly from IR studies. In traditional approaches, in general, a document is represented as a vector in a feature space formed by all the keywords [24]. Typically, people use either Cosine-similarity, Jaccard-similarity or Dice-similarity [24] as similarity functions. Keywords are all the words, except the function words included in a stop-list. All the keywords can be stemmed using the Porter algorithm [21]. The keyword-based similarity function is defined as follows:

$$similarity_{keyword}(q_1, q_2) = \frac{KN(q_1, q_2)}{max(kn(q_1), kn(q_2))} \quad (1)$$

where $kn(.)$ is the number of the keywords in a query, $KN(q_1, q_2)$ is the number of common keywords in two queries.

If query terms are weighted, the Cosine similarity [24] can be used instead:

$$similarity_{weighted\text{-}keyword}(q_1, q_2) = \frac{\sum_{i=1}^{k} cw_i(q_1) \times cw_i(q_2)}{\sqrt{\sum_{i=1}^{m} w_i^2(q_1)} \times \sqrt{\sum_{i=1}^{n} w_i^2(q_2)}} \quad (2)$$

where $cw_i(q_1)$ and $cw_i(q_2)$ are the weights of the *i-th* common keyword in the query q_1 and q_2 respectively, and $w_i(q_1)$ and $w_i(q_2)$ are the weights of the *i-th* keywords in the query q_1 and q_2 respectively. Usually, $tf * idf$ [24] is used for keyword weighting.

The above measures can be easily extended to phrases. Since phrases are a more precise representation of meaning than single words, if we can identify phrases in queries, we can obtain a more accurate calculation of query similarity. For example, the two queries "history of China" and "history of the United States" are very close queries (both asking about the history of a country). Their similarity is 0.33 on the basis of keywords. If we can recognize "the United States" as a phrase and take it as a single term, the similarity between these two queries is increased to 0.5. The calculation of phrase-based similarity is similar to Formula (1) and (2). We only need to recognize phrases in queries. There are two main methods for doing this. One is by using a noun phrase recognizer based on syntactic rules and statistics [4][15]. Another way is to use a phrase dictionary. A phrase dictionary can be constructed manually or using an automatic phrase recognizer based on a syntactic and statistical analysis on corpus.

The keyword-based measure has provided interesting results in the document clustering domain. One contributing factor is the large number of keywords contained in documents. Even if some of the keywords of two similar documents are different, there are still many other keywords that can make the documents similar in the similarity calculation. However, since queries, especially the queries submitted to the search engines, are typically very short, in many cases it is hard to deduce the semantics from the queries themselves. Therefore, keywords alone do not provide a reliable basis for clustering queries effectively.

2.2 Similarity Based on Word-Level String Matching

Keyword-based measure treats words such as "where", "who" and "why" as stop words. For questions, however, these words (if they occur) encode important information about the user's need, particularly in the question-answering search engines such as AskJeeves. For example, with a "who"-question, the user intends to find information about a person. So, even if a keyword-based approach is used in query clustering, it should be modified from that used in traditional document clustering. Special attention is paid to such words in Question Answering (QA) [14][25], where they are used as prominent indicators of question type. The whole question is represented as a template in accordance with the question type. During question evaluation, the question template may be expanded using a thesaurus (e.g. WordNet [17]) or morphological transformations.

In the web context, well-formed natural language questions represented only a small portion of queries. Most queries are simply short phrases or keywords (e.g. "population of U.S."). The approach used in QA is therefore not completely applicable to the web query clustering problem. However, if words denoting a question type do appear in a complete question, these words should be taken into account.

Here we define a similarity measure that uses all the words in the queries for similarity estimation, even the stop words. Comparison between queries becomes an inexact string-matching problem, as formulated by [10]. Similarity may be determined by the edit distance, which is a measure based on the number of edit operations (insertion, deletion, or substitution of a word) necessary to unify two strings (queries). The maximum number of the words in the two queries is used to divide the edit distance so that the value can be constrained within the range of $[0, 1]$. The similarity is inversely proportional to the normalized edit distance:

$$similarity_{edit}(q_1, q_2) = 1 - \frac{edit\text{-}distance(q_1, q_2)}{max(wn(q_1), wn(q_2))} \tag{3}$$

where $wn(.)$ is the number of the words in a query.

The advantage of this measure is that it takes into account the word order, as well as words that denote query types such as "who" and "what" if they appear in a query. This method is more flexible than those used in QA systems, which rely on special recognition mechanisms for different types of questions. It is therefore more suitable to the query clustering situation. This measure is particularly useful for long and complete questions in natural language. Below are some questions that have been grouped into one cluster:

- Query 1: How far away is the moon?
- Query 2: How far is the closest star?
- Query 3: How far is the nearest star?

This cluster contains questions of the form "How far (away) is the X".

In the similarity calculations described above, we can further incorporate a dictionary of synonyms. If two words/terms are synonyms, their similarity is set at a predetermined value (e.g., 0.8). It is then easy to incorporate this similarity between synonyms into Formula (1), (2) and (3).

2.3 Similarity Based on Character-Level String Matching

The web user often misspells his/her query because the user does not know the spelling, or because of a typing error, or because the user is not completely sure about the spelling. According to [13], the four error types, insertion, deletion, substitution and transposition encompass 80 percent of all spelling errors. Some typical errors like misspelling "Pokemon" as "Pokeman" occur very frequently when users using search engines. Query clustering can also be used to detect and correct such a kind of spelling errors in user queries.

The word-level string matching method above is incapable of dealing with the spelling error problem. Instead, just like [13], the character-level string matching method is effective to group a query with its misspelling counterparts. In this case, the edit distance in Formula (3) is defined as the number of edit operations (insertion, deletion, or substitution of a character) necessary to unify two queries.

Normally a search engine does not give any answers to queries when the search words do not exist in the indices. Once we can group the queries with spelling errors together, we find a way to collect the most frequent spelling errors in user queries. Then an automatic query correction mechanism could be employed to help the users search the Web more smoothly, such as the *searchspell* website (http://www.searchspell.com/typo). Although query clustering by itself cannot solve the misspelling problem thoroughly, it does provide the website editors a powerful tool to assist them to detect and correct spelling errors more easily.

3 Session-based Query Clustering

Due to the short length and word ambiguity of user queries, content-based clustering methods usually cannot group those semantically related queries. Session-based query clustering is a newly emerging strategy, which employs the evidences in query sessions to deduce and detect users' search intentions. Many search engines have accumulated a large amount of web query logs, from which we can find out what the query is, and the web pages the user has selected to browse. Typically, a query session is made up of the query and the subsequent activities the user performed, which can be extracted from query logs. Instead of relying only on the query terms themselves, session-based query clustering methods take advantage of various evidences hidden in query sessions to calculate the similarity between queries.

One important assumption behind these methods is that the activities following a query are relevant to the query. User activities can be viewed as (implicit) user relevance feedback in a traditional IR environment. At first glance, this assumption may seem to be too strong. Users may click the irrelevant web pages, users may make error-clicks, and users may shift their interests after submitting the query. However, in the web environment, the choice of a particular web page from a list by a user is a good indication of the user's search intention. Although it is not as accurate as explicit relevance judgment in traditional IR, the user's choice does suggest a certain degree of "relevance" of that web page to his information need. Users usually have rough ideas about what they want when submitting queries. So when they get a list of web pages, users do not select them randomly. Instead, they click and read those web pages which are the most similar to what are in their minds based on the information provided by the search engine on the returned candidate answers (e.g. titles and snippets). Therefore, these clicked web pages do have some relationship with the queries they submit. Even if some of the web page clicks are erroneous, we can expect that most users do click on web pages that are relevant.

In fact, using cross-reference between queries and user activities to improve retrieval performance is not a new idea in the IR field. Relevance feedback in IR is a typical exploitation of cross-references [22][23], which modifies a query based on user's relevance judgments of the retrieved documents. Relevance feedback can achieve very good performance if the users provide sufficient and correct relevance judgments. Unfortunately, in a real search context, users usually are reluctant to provide such relevance feedback information. Therefore, relevance feedback is seldom used by the commercial search engines. To overcome the difficulty due to the lack of sufficient relevance judgments, pseudo-relevance feedback (also known as blind feedback) is commonly used. Local feedback mimics relevance feedback by assuming the top-ranked documents to be relevant [31]. Nevertheless, this method has an obvious drawback: if a large fraction of the top-ranked documents is actually irrelevant, the performance of this method is likely to be worse. Now lets look at query logs. Query logs can be taken as a very valuable resource containing abundant relevance feedback data, which is not available in the traditional IR environment. With the query logs, we can overcome the problem of lacking sufficient relevance judgments in traditional relevance feedback technique. In comparison with pseudo-relevance feedback, the query session based method has an obvious advantage: not only are the clicked web pages part of the top-ranked web pages, but also there is a further selection by the user. So clicks are more reliable indications than those used in pseudo relevance feedback.

Similar ideas have been used in some other work in IR. Voorhees et al. [27] try to evaluate the quality of an IR system for a cluster of queries on different document collections. Their goal is to determine an appropriate method for database merging according to the quality estimation. The assumption used in query clustering is that if two queries retrieve many documents in common they are on the same topic. In [16], a measure of query similarity is used to route queries to proper collection replicas. In their work, two queries are considered to be related to the same topic if the top 20 documents they retrieve completely overlap.

3.1 Query Session Extraction

Typically, web logs will track the activities of the users who are utilizing a search engine. Each record in the web log includes the IP address of the client computer, timestamp, the URL of the requested item, type of web browser, protocol, etc. The web log of a search engine records various kinds of user activities, such as getting HTML pages, getting images and running Java scripts. Although all these activities reflect, more or less, a user's intention, the query terms and the web pages the user visited are the most solid evidences to guess users' search intention. Therefore, a query session is the query a user submitted to a search engine together with the web pages he/she visits in response to his/her request. Table 1 lists a small excerpt of query log from the Encarta web site (We omit some fields such as the type of web browser, the IP address of the server, the bytes sent or received, etc.). This query session shows that a user submitted a query "atomic bomb" to the search engine and then browsed two web pages, which are two Encarta articles related to atomic bomb.

Since the HTTP protocol requires a separate connection for every client-server interaction, the activities of multiple users usually interleaved with each other. There are no clear boundaries among user query sessions in the logs, which makes it a difficult task to extract individual query sessions from web query logs, as addressed in [2][11][20].

Query session extraction mainly contains two steps — user identification and session identification. User identification is the process of isolating from the logs the activities associated with an individual user. The most intuitive way of user identification is to group the activities with the same IP address. But IP address is not a reliable indication to distinguish users in the case of the broad existence of local caches, corporate firewalls, and proxy servers. In a web server log, all requests from a proxy server have the same identifier, even though the requests are potentially from more than one user. Also, due to proxy server level caching, a single request from the server could actually be viewed by multiple users. [2][19] proposed some heuristics to distinguish different users with the same IP address. For example, if the agent log shows a change in browser software or operating system, a reasonable assumption to make is that each different agent type for an IP address represents a different user. Another heuristic for user identification is to use the access log in conjunction with site topology to construct browsing paths for each user. That is if a requested page is not directly reachable by a hyperlink from any of the pages visited by the user, the heuristic assumes that there is another user with the same IP address. However, these heuristics are not "perfect" solution to the user identification problem. Two users with the same IP address that use the same browser on the same type of machine can easily be confused as a single user if they are looking at the same set of pages. Cookies are markers that are used to tag and track site visitors automatically. Cookies are more reliable than IP addresses because, in general, the cookies from two computers are different, even when they use the same IP address. But cookies can be deleted by the user since he/she doesn't want the server to track his/her behaviors. Another method to identify users is through user ID. User ID has the advantage of being able to collect additional demographic information beyond what is automatically collected in the server log, as well as simplifying the identification of user sessions. However, again due to privacy concerns, many users choose not to browse sites that require registration and logins. There is a constant struggle between the web user's

Table 1: Sample Query Session

IP address	Cookie	Timestamp	URL
999.999.999.1	FF4AF1D6935A	4/25/00:2:14:14	http://encarta.msn.com/ find/search.asp?s=atomic+bomb
999.999.999.1	FF4AF1D6935A	4/25/00:2:14:38	http://encarta.msn.com/ article/05808000.html
999.999.999.1	FF4AF1D6935A	4/25/00:2:21:45	http://encarta.msn.com/ article/01FC9000.html

desire for privacy and the web provider's desire for information about who is using their site. Many users choose to disable the browser features that enable these methods.

The goal of session identification is to divide the queries and page accesses of each user into individual sessions. Finding the beginning of a query session is trivial: a query session begins when a user submits a search to a search engine. However, it is more difficult to determine when a search session ends. The simplest method of achieving this is through a timeout, where if the time between page requests exceeds a certain limit, it is assumed that the user is starting a new session. Many commercial products use 30 minutes as a default timeout. Another way is to get the timeout through analyzing empirical data. Once a site log has been analyzed and usage statistics obtained, a timeout that is appropriate for the specific Web site can be fed back into the session identification algorithm.

3.2 Naive Similarity Based on Clickthroughs

The first session-based similarity function is to simply use clicked pages as the semantic descriptions of the user queries and considers each web page in isolation. This query clustering approach is based on the following principle: *If users clicked on the same web pages for different queries, then these queries are similar.* The formal description of this similarity function is as follows:

Let $D(q_1)$ and $D(q_2)$ be the set of web pages the system presents to the user as search the results for the queries q_1 and q_2 respectively. The page set that users clicked on for the queries q_1 and q_2 may be seen as:

$$DC(q_1) = \{d_{q_{11}}, d_{q_{12}}, \ldots, d_{q_{1i}}\} \subseteq D(q_1)$$
$$DC(q_2) = \{d_{q_{21}}, d_{q_{22}}, \ldots, d_{q_{2j}}\} \subseteq D(q_2)$$

Similarity based on clickthroughs follows the following principle: If $DC(q_1) \cap DC(q_2) = \{d_1, d_2, \ldots, d_k\} \neq \emptyset$, then web pages $d_1, d_2, \ldots, d_k$ represent the common topics of queries q_1 and q_2. The similarity between the queries q_1 and q_2 is determined by $DC(q_1) \cap DC(q_2)$. Therefore, the similarity between the two queries is proportional to the shared number of clicked (or selected) web pages, taken individually, as follows:

$$similarity_{naive\text{-}session}(q_1, q_2) = \frac{DN(q_1, q_2)}{max(dn(q_1), dn(q_2))} \quad (4)$$

where $dn(.)$ is the number of clicked pages for a query, $DN(q_1, q_2)$ is the number of page clicks in common.

In spite of its simplicity, this measure demonstrates a surprising ability to cluster semantically related queries that contain different words. In Figure 1, "Atomic bomb", "Nagasaki", and "Nuclear weapon" are clustered as similar queries since all of them correspond to a clicked web page about atomic bomb. It is obvious that no approach that uses only the keywords in the queries could create such a cluster since these queries are made up of different keywords.

In addition, this measure is also very useful in distinguishing between queries that happen to be worded similarly but stem from different information needs. For example, in Figure 1, one user asked for "java" and clicked on the web pages about tour in Java Island, one user asked "java" and clicked the web pages about Java programming language, and another user asked the same "java" query and clicked the web pages about Java coffee. These different search intentions can be easily distinguished through user clicks. But the content-based method will certainly consider all of them as the same query. This kind of distinction can be further exploited for sense disambiguation in a user interface.

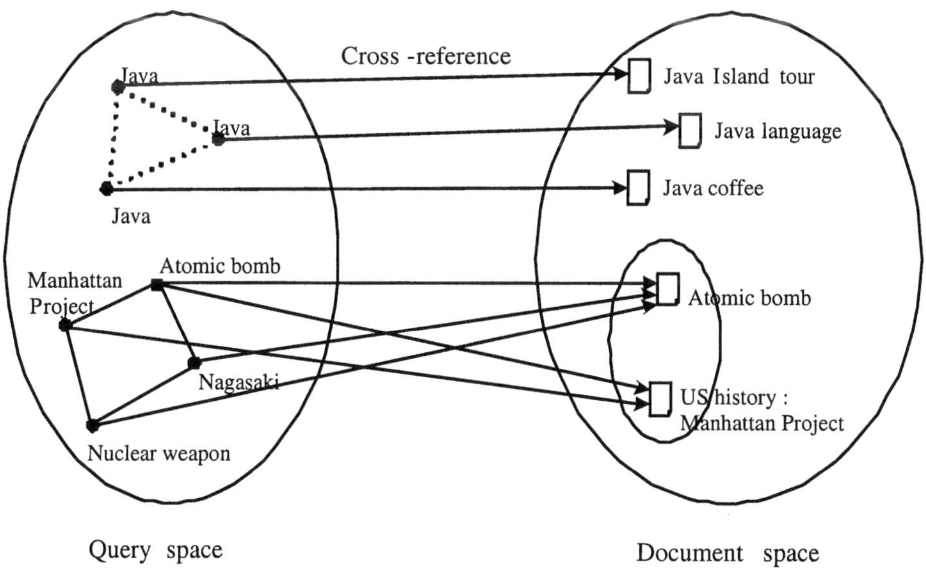

Figure 1: Session-based Query Clustering

3.3 Combination of Query Clustering and Document Clustering

The above similarity function is a naive one and only effective when the clicked web pages are constrained in a relatively narrow range and many users click the same web pages. Especially, when two queries leading to two separate but similar page clicks, the similarity between these two queries cannot be detected. In Figure 1, "Manhattan Project" is not detected as a similar query to "atomic bomb" because its related web pages are about "US history: Manhattan Project", although they are very close to the "atomic bomb" web page but not exactly the same.

To overcome this difficulty and obtain query clusters with broader coverage, document clustering can be used to group the similar web pages first, and then two queries pointing to two different web pages in the same cluster can be recognized as similar queries. There are two ways to achieve such a kind of combination of query clustering and document clustering — unidirectional clustering and bidirectional clustering. In unidirectional clustering, documents need to be collected and clustered ahead of time, and then the document clusters are used to calculate the similarities between queries. In this case, the document clustering and query clustering are two separate phases. On the contrary, bidirectional clustering employs an iterative, agglomerative method to cluster queries and documents simultaneously.

3.3.1 Unidirectional Clustering

Document clustering has been investigated for use in a number of different areas of text mining and information retrieval for decades. Numerous document clustering algorithms have appeared in the literature [30]. In recent years, document clustering was also extended to organizing the web pages returned by a search engine [32].

There are two ways to take advantage of document clustering. One of them is relatively straightforward. First, all web pages are scattered into n different clusters $C_1, C_2, \ldots, C_n$, using any document clustering algorithm. For each query session, all clicked web pages are substituted by the clusters they belong to. Thus, the similarity between two queries is proportional to their shared number of clusters. Then Formula (4) is used to calculate the similarity value between the two queries, except that $DN(q_1, q_2)$ is the number of clicked clusters in common. For the example in Figure 1, if web pages "atomic bomb" and "US history: Manhattan Project" are grouped into a cluster, queries "atomic bomb" and "Manhattan Project" will be considered as leading to the same clicks and put into the same cluster.

The other way to make use of document clustering is to directly incorporate document similarity measure into the calculation of query similarity. Let $s(d_i, d_j)$ be the function to calculate the similarity between documents, $d_i (1 \leq i \leq m)$ and $d_j (1 \leq j \leq n)$ be the clicked documents for queries q_1 and q_2 respectively. The incorporated similarity function is defined as follows:

$$similarity_{session+doc}(q_1, q_2) = \frac{1}{2} \times \left(\frac{\sum_{i=1}^{m}(max_{j=1}^{n} s(d_i, d_j))}{m} + \right. \\ \left. + \frac{\sum_{j=1}^{n}(max_{i=1}^{m} s(d_i, d_j))}{n}\right) \qquad (5)$$

For the same example in Figure 1, if the similarity value between the two web pages "atomic bomb" and "US history: Manhattan Project" is 0.5, and they are the only two web pages selected for the two queries, then the similarity value between the queries is also 0.5 according to Formula (5). In contrast, their similarity value based on Formula (4) using common clicks is 0. Hence, this new similarity function can recognize a wider range of similar queries.

3.3.2 Bidirectional Clustering

As illustrated above, document clustering may effectively enhance query clustering unilaterally. But clustering web pages by content may require storing and manipulating quite a large amount of data, and the unidirectional strategy is only effective when the web pages are constrained in a relatively narrow range. Therefore, cross-reference between queries and web pages is also used to cluster similar documents in a "content-ignorance" manner, that is, *if many people select a common set of documents for the same query, this is a strong indication that this set of documents is similar.* This is a dual hypothesis of the session-based query clustering hypothesis, and has been used in several existing search engines. For example, in Citeseer (http://citeseer.nj.nec.com), once a document is identified, the system also indicates a set of other documents that other people have accessed together with that document. This is a form of implicit document clustering based on cross-references using the above hypothesis. Amazon (http://amazon.com) exploits the same hypothesis to suggest books similar to those selected by users.

In summary, through query sessions, the distance between two queries can be evaluated by examining their corresponding clicked web pages, and the distance between two web pages can be evaluated by examining their related queries. Therefore, query clustering and (web) document clustering procedures can be combined and reinforced each other to construct a bidirectional clustering method. The intuition of the bidirectional clustering is that similarity between documents can be transferred to queries through their cross-references, and vice versa. Below we discuss a bidirectional clustering algorithm introduced in [1].

The work in [1] is conducted on the Lycos website. First, based on query sessions in web logs, a bipartite graph is constructed where the vertices on one side correspond to unique queries, the vertices on the other side to unique web pages, and edges exist between a query and a web page vertex when they co-occurred in a query session. Figure 2(a) is a sample bipartite graph. Second, a similarity function is defined to measure the similarity between vertices x and y, both for query vertices and web page vertices. Then an agglomerative clustering algorithm is applied to cluster the queries and web pages respectively. The iterations of these two agglomerative clustering procedures are not independent but interleave each other — first combing the two most similar queries, then combining the two most similar web pages, and repeat the iterations until the resulting graph consists of connected components, each with a single query and web page vertex. Figure 2(b) is the graph at the termination state.

This bidirectional clustering strategy is very suitable for clustering queries and web pages simultaneously on general-domain search engine logs with sparse clickthroughs on a huge collection of diverse web pages. The "content-ignorance" characteristic is very valuable when dealing with large datasets, dynamic web pages, text-free pages and pages with restricted access. But while reinforcing query clustering and document clustering each other, bidirectional clustering is also likely to accumulate noises and outliers for both sides. Another unsolved problem is that it is hard to determine the termination condition in advance and the condition in [1] is obviously too severe. Moreover, while exploiting cross-references, [1] rejects the use of the contents of queries and documents. They consider that keyword similarity is totally unreliable. We believe that content words provide some useful information for query clustering that is complementary to cross-references.

3.4 Clustering Queries through Query Co-occurrences

In this part, we briefly introduce another viewpoint about query session and query clustering. As stated in [12], a series of related queries of an individual users can be extracted from web logs and to form a new kind of query session, as shown in Table 2. The session in Table 2 illustrates

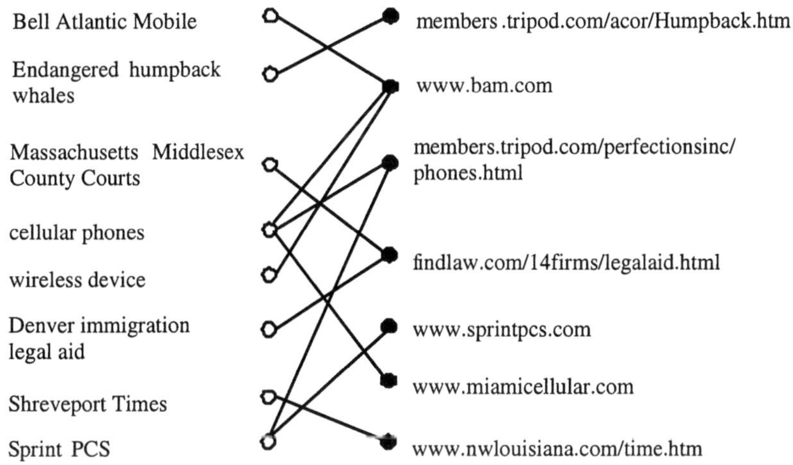

(a) Original Bipartite Graph

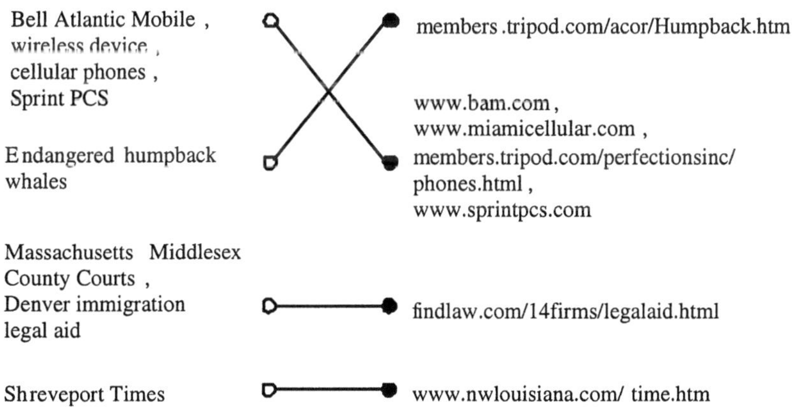

(b) Graph at the Termination State

Figure 2: Bidirectional Clustering

a typical search scenario: the user first submitted query "WW2" to the search engine and looked for information about the World War Two. Then he/she refined his/her query to "second war" (maybe due to not finding satisfactory answers). Finally, the user submitted the query "world war 2" and stopped this search session. From this scenario we can see that the query co-occurrences in a session can be taken as a useful evidence to judge the correlations among queries [12][8].

Table 2: Another Kind of Query Session

IP address	Cookie	Timestamp	URL
999.999.999.2	BCA8D4C3778E	5/12/00:6:23:21	http://encarta.msn.com/ find/search.asp?s=WW2
999.999.999.2	BCA8D4C3778E	5/12/00:6:23:52	http://encarta.msn.com/ find/search.asp?s=second+war
999.999.999.2	BCA8D4C3778E	5/12/00:6:28:06	http://encarta.msn.com/ find/search.asp?s=world+war+2

Although the above example happens frequently in web logs, another common scenario is that consecutive queries from a user are independent from each other and the user actually changed his/her interests during the search process. The main challenge is that it's very hard to partition the query log into query sessions so that each query session really corresponds to the query terms submitted by a user for a single information need. Another problem is how to measure the relevance between each pair of query terms in a session.

4 Combining Evidences from Query Content and Query Session

As we mentioned before, content-based similarity function alone is not sufficient to cluster semantically related queries. The main reason is that keywords and meanings do not strictly correspond. The same keyword does not always represent the same information need (e.g. the word "Java" may refer to "Java" the island, coffee, or a programming language); and different keywords may refer to the same concept (e.g. all the terms "atomic bomb", "Manhattan project", "nuclear weapon" and "Nagasaki" are related to the concept "atomic bomb"). Therefore, calculated similarity does not necessarily reflect semantic similarity. This is particularly the case for web queries which typically are very short (usually less than two words [29]).

On the other hand, one may reject the use of query sessions as a reliable criterion, too. In fact, a web page click is not a relevance judgment. Users may well click on irrelevant pages. Therefore, web page clicks, if considered as relevance judgments, are noisy. In addition, a web page can describe more than one specific topic, and can correspond to different queries. Putting together the queries corresponding to the same web page may result in a cluster containing very different queries. These facts may raise reasonable doubt on the reliability of web page clicks as a clustering criterion.

While it is true that relations made by a small number of users are likely to be unreliable, the large amount of information available in web logs makes this less of a problem; we assume that users are more consistent in their relations of relevant web pages than irrelevant ones. Cross-references confirmed by many users are therefore taken to be reliable. One more supporting evidence is the results of pseudo-relevance feedback in IR [31] which show that in reformulating queries using the top-ranked documents as if they are relevant, the retrieval effectiveness can be increased significantly. This shows that useful information about the connection between queries and documents in IR is not restricted to explicit relevance judgments. Even very coarse and noisy indications can help. Our situation is better than in pseudo-relevance feedback — not only are the documents presented to the user the top-ranked documents, but there is also a further selection by the user. So document clicks are more reliable indication than that used in pseudo-relevance feedback. Therefore, we can only expect better results.

Examinations of the typical web logs showed that most of the queries are followed by only one or two web page selections. People may doubt whether such a small number of clicks can provide a reliable basis for query clustering. This is very similar to the "short-query" problem. Both content words and web page clicks can partially reflect the semantics of the queries. But because of the small numbers of keywords and document clicks, neither of them is sufficient to be used alone. Therefore, some combined measures should be defined to take advantage of both strategies. A simple way to do this is to combine both measures linearly, as follows:

$$similarity = \alpha \times similarity_{content\text{-}based} + \beta \times similarity_{session\text{-}based} \quad (6)$$

Some experiments [29] show that the combined measure can be used to cluster more queries in a more precise way. The combined measure raises the question of how to set parameters α and β. We believe that these parameters should be set according to the editor's objectives.

5 Choosing a Clustering Algorithm

Besides similarity functions, choosing an appropriate clustering algorithm is also very critical to the effectiveness and efficiency of the query clustering process. As to choosing web query clustering algorithm, there are several rules that should be followed. First, as query logs usually are very large, the algorithm should be capable of handling a large data set within reasonable time and space constraints. Second, due to the fact that the log data changes daily, the algorithm should be easily extended to cluster new queries incrementally. Third, the algorithm should not require manual setting of the resulting form of the clusters, e.g. the number or the maximal size of clusters. It is unreasonable to determine these parameters in advance. Finally, since we only want to find FAQs, the algorithm should filter out those queries with low frequencies.

There is a great deal of work done previously in clustering algorithms. Basically, they can be divided into two groups of clustering algorithms: non-hierarchical and hierarchical. K-Means and Agglomerative Hierarchical Clustering (AHC) are representatives of these two groups [5].

K-Means constructs a partition of n points into the final required k clusters. First, k clusters is chosen such that the points are mutually farthest apart. Next, it examines each point and assigns it to one of the clusters depending on the minimum distance. The centroid of a cluster is recalculated every time a point is added to it and this continues until all the points are grouped into the final required number of clusters. K-Means has a time complexity which is linear in the number of points, which complies with the speed requirement of clustering a large amount of queries. Its main disadvantage is that it is known to be the most effective when the desired clusters are approximately spherical with respect to the similarity measure used. There is no reason to believe that queries should fall into approximately spherical clusters. Another disadvantage of K-Means is that the number of clusters, k, needs to be determined at the onset, which is usually unrealistic.

Agglomerative Hierarchical Clustering is probably the most commonly used clustering algorithm and it creates a hierarchical decomposition of a population of points. In AHC, each object is initially placed into its own group. Therefore, if we have n objects to cluster, we start with n groups. Each of these groups contains only a single object and is known as a singleton. If the distance between a closest pair of groups is less than the threshold distance, these groups become linked and are merged into a single group. This merge procedure is repeated until there is no other merge possible. To run an agglomerative clustering, a measure should be provided to calculate the distance between two objects. In addition, another measure is needed to determine which groups should be linked. Some options are single-link, group-average, complete-link, etc. AHC is typically slow when applied to large dataset. Single-link and group-average methods typically take $O(n^2)$ time, while complete-link methods typically take $O(n^3)$ time. Although it is not required to know the number of final clusters ahead of time, the halting criteria of AHC are arbitrary and its performance is poor in domains with many outliers.

The density-based clustering method, DBSCAN [6] and its incremental version, Incremental DBSCAN [7] are good candidate algorithms which satisfy all of the above requirements. DBSCAN does not require the number of clusters as an aprior input parameter. Density-based methods are based on the idea that it is likely that in a space of objects, dense objects should be grouped together into one cluster. Thus, a cluster is a region that has a higher density of points than its surrounding region. A cluster consists of at least the minimum number of points — *MinPts* (to eliminate very small clusters as noise); and for every point in the cluster, there exists another point in the same cluster whose distance is less than the distance threshold *Eps* (points are densely located). DBSCAN is a one pass algorithm since it groups neighboring objects of the database into clusters based on a local cluster condition. It is very efficient if the retrieval of the neighborhood of an object is efficiently supported by some indexing mechanism. For the query clustering case, an inverted file is very efficient to search for neighbor points. All clusters consisting of less than the minimum number of points are considered as "noise" and are discarded. Thus those queries with low frequencies will be filter out. Previous experiments showed that DBSCAN outperforms CLARANS [18] by a factor of between 250 and 1900, which increases with the size of the data set. The ability of Incremental DBSCAN to update incrementally is due to the density-based nature of the DBSCAN method, in which the insertion or deletion of an object only affects the neighborhood of this object. In addition, based on the formal definition of clusters, it has been proven that the incremental algorithm yields the same results as DBSCAN. Performance evaluations show that Incremental DBSCAN is more efficient than the basic DBSCAN algorithm.

6 A Case Study on Encarta Online Search Engine

To test various query clustering methods, we carried out a study on the Encarta encyclopedia (http://encarta.msn.com), which can be accessed on the Web. The Encarta editors have an impending need of determining the FAQs of the users. This research work is launched specifically to answer the need of Encarta editors.

Encarta contains a large number of documents organized in a hierarchy. A document (or article) is written on a specific topic, for instance, about a country. In contrast with other documents on the Web, the quality of the documents is well controlled. The current search engine used on Encarta uses some advanced strategies to retrieve relevant documents using keywords. The answer to a query/question is a set of documents. Although the search engine suffers from the same common problems of search engines using keywords, the result list does provide a good indication of the document contents. Therefore, when a user clicks on one of the documents in the result list, he/she has some idea about what can be found in the document. This provides us with a solid basis for using user clicks as a valid indication of relevance.

Encarta online is not a static encyclopedia as the printed one. It evolves in time. In fact, a number of human editors and indexers are working on the improvement of the encyclopedia so that users can find more information from Encarta and in a more precise way. Human editors work on expanding contents by adding new documents according to users' "hot topics". And human indexers want to increase search performance by adding new index entries or adjusting index term weights according to users' "hot query terms". Moreover, if many users asked the same questions (FAQ) in a certain period of time, then the answers to these questions will be checked manually and directly linked to the questions.

For a long time, all of these tasks are done based on manual analysis of user queries in the logs. Our goal is to develop a clustering system that helps the editors quickly identify the FAQs. The key problem is to determine a similarity function which will enable the system to form clusters of truly similar queries.

6.1 Query Sessions

We collected one-month user logs (about 22 GB) from the Encarta web site. From these logs we extracted more than two millions user query sessions. A query session is defined as: $session = query \rightarrow [clicked\text{-}document]^*$. Each session corresponds to one query and the Encarta documents the user clicked on. 49% of queries contain only one keyword and 33% of queries contain two keywords. The average length of all queries is 1.86. The distribution of query length is similar to those reported by others. Because the number of queries is too big to conduct detailed evaluations, we randomly chose 20,000 query sessions from them for our study.

In Encarta, once a user query is input, a list of documents is presented to the user, together with the document titles and snippets. We observed that about half of query sessions have document clicks. Out of these sessions, about 90% have 1–2 document clicks. Because the document titles and snippets in Encarta are carefully chosen, they give the user a good idea of the contents of the documents. Therefore, if a user clicks on a document, it is likely that the document is relevant to the query, or at least related to it to some extent. Even if some of the document clicks are erroneous, we can expect that most users do click on relevant/strongly related documents.

6.2 Document Collection and Document Similarity

The Encarta documents collection is made up of 41,942 documents in various topics. Those documents are not isolated; they are organized into a hierarchy that corresponds to a concept space. The hierarchy contains 4 levels. The first level is the root. The second level contains 9 categories, such as "physical science & technology", "life science", "geography", etc. These categories are divided into 93 subcategories. The last level (the leaf nodes) is made up of tens of thousands of documents. Roughly, this document hierarchy corresponds to a domain/concept hierarchy. Instead of using document contents, we use the document hierarchy to calculate conceptual distances between documents. This distance is determined as follows: the lower the common parent node the two documents have, the shorter the conceptual distance between the two documents is, and the higher their conceptual similarity is. Let $F(d_i, d_j)$ denote the lowest common parent node for documents d_i and d_j, $L(x)$ the level of node x, *L-Total* the total levels in the hierarchy (i.e. 4 for Encarta). The conceptual similarity between two documents is defined as follows:

$$s(d_i, d_j) = \frac{L(F(d_i, d_j)) - 1}{L\text{-}Total - 1} \tag{7}$$

In particular, $s(d_i, d_i) = 1$; and $s(d_i, d_j) = 0$ if $F(d_i, d_j) = root$.

We incorporate this document similarity measure into Formula (5) and use it as a new similarity measure for queries.

6.3 Evaluation of Different Similarity Functions

To test how different similarity functions affect the query clustering results, we conducted a series of experiments using Encarta logs and document collection. Specifically, we tested the following four similarity functions on the extracted 20,000 query sessions:

- Query keyword similarity (*K-Sim, Formula (1)*),
- Naive similarity using clickthroughs (*S-Sim, Formula (4)*),
- Query keyword + Naive clickthrough similarity (*K+S-Sim, Formula (1)+(4)*), and
- Query keyword + clickthrough + document similarity (*K+S+D-Sim, Formula (1)+(5)+(7)*).

DBSCAN is chosen as the clustering algorithm. The minimal density parameter ($MinPts$) was set to 3 uniformly, which means that only those clusters containing at least 3 queries are kept. Then we varied the similarity threshold ($= 1 - Eps$) from 0.5 to 1.0. As for the combined measures, we assigned weight 0.5 to both α and β uniformly.

Like other applications using clustering algorithms, there is no standard methodology to evaluate the quality of the results of query clustering. We borrow two IR metrics as our quality measures:

- Precision — the ratio of the number of similar queries to the total number of queries in a cluster.
- Recall — the ratio of the number of similar queries to the total number of all similar queries for these queries (those in the current cluster and others).

For every similarity function, we randomly selected 100 clusters. For each cluster, we calculated precision by manually checking the queries. Since we do not know the actual intentions of users with their queries, we simply guess, taking into account both queries and clicked documents. We report the average precision of the 100 clusters in Figure 3, where all the four functions are shown with similarity threshold varying from 0.5 to 1.0.

We first observe that the combinations of keywords and clickthroughs in query sessions (*K+S-Sim* and *K+S+D-Sim*) result in higher precision than the two criteria separately. When similarity threshold is equal to or higher than 0.6, *K+S-Sim* and *K+S+D-Sim* have very high precision (above 95%). When the similarity threshold is higher than 0.8, the precision for both similarity functions reaches 1.0.

For clustering using single evidence, we observe that the highest precision that can be reached by *K-Sim* is about 96%, when all queries in a cluster contain identical keywords. This means the ambiguity of keywords will only bring in about 4% errors. This number is much lower than our expectation. A possible reason is that users usually are aware of word ambiguity and would like to use more precise queries. For example, instead of using "Java", users use "Java Island" or "Java programming language" to avoid ambiguities.

It is difficult to use the recall metric directly for clustering because no standard clusters or classes are available. Therefore, we use a different measure to reflect, to some extent, the recall factor: normalized recall. This factor is calculated as follows:

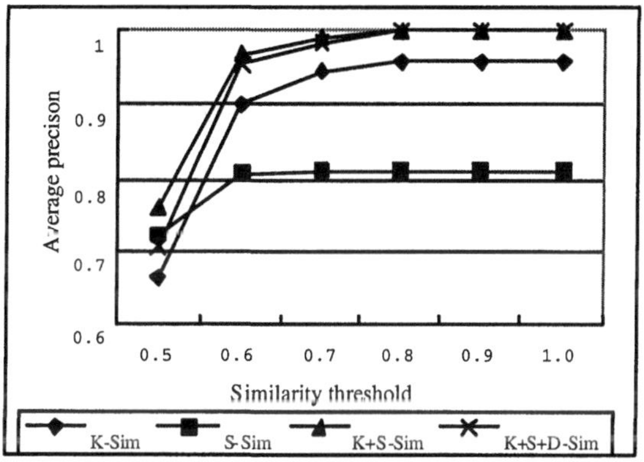

Figure 3: Precision for Four Kinds of Similarity Functions

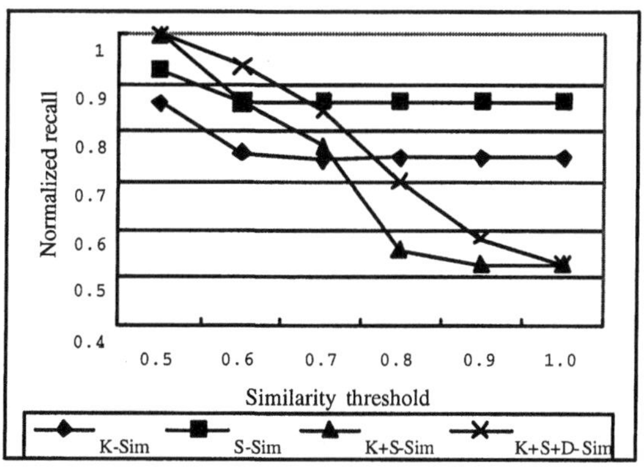

Figure 4: Recall for Four Kinds of Similarity Functions

- For any similarity function, we collect the number of correctly clustered queries in all the 100 clusters. This is equal to the total number of clustered queries times the precision.

- Then we normalize this value by dividing it with the maximum number of correctly clustered queries. In our case, this number is 12357 which is obtained by *K+S+D-Sim* when the similarity threshold is 0.5. This normalization aims at obtaining a number in the range [0, 1].

Figure 4 shows the normalized recalls for the four similarity functions when similarity threshold varies from 0.5 to 1.0. We observe that, when similarity threshold is below 0.6, *K+S+D-Sim* and *K+S-Sim* result in better normalized recall ratios than using two other functions on single evidence. This shows that both functions can take advantage of evidences from query and session. However, when similarity threshold increases, then normalized recall ratios drop quickly. On the other hand, there is almost no change for *S-Sim* and *K-Sim* for threshold higher than 0.6. Again, this is due to the small number of keywords per query and document clicks per query.

Although the precision of *K+S+D-Sim* is very close to *K+S-Sim* (Figure 3), the normalized recall of *K+S+D-Sim* is always higher than *K+S-Sim* with a margin of about 10% (Figure 4). This shows that, when combined with keywords, the consideration of document similarity is helpful for increasing recall significantly without decreasing precision.

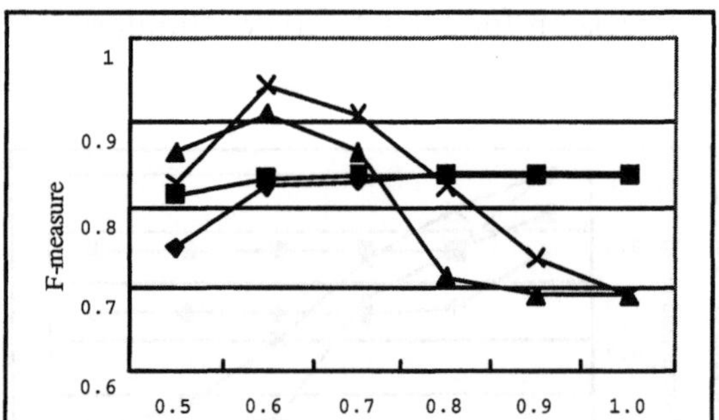

Figure 5: F-measures for Four Kinds of Similarity Functions

In order to compare the global quality of the four functions, we use the F-measure [16] as metric (in which the recall ratio is replaced by our normalized recall). Although the modified F-measure is different from the traditional one, it does provide some indication on the global quality of different similarity functions. Figure 5 shows this evaluation. We can see that within the threshold range of $[0.5, 0.7]$, *K+S+D-Sim* and *K+S-Sim* are better than the single-evidence functions. In particular, when similarity threshold is equal to 0.6, *K+S+D-Sim* reaches the highest F-measure value (0.94).

All the above experiments show that it is beneficial to combine evidences from query contents, query sessions, and documents in query clustering.

7 Summary

Query clustering is a class of techniques developed in recent years to meet the requirements of more sophisticated web search. Query clustering is not an easy task. It requires a proper estimation of query similarity. Unlike traditional document clustering, due to the short length and word ambiguity of user queries, the similarity between queries cannot be accurately measured only based on query contents (e.g., keywords, query forms, etc.) and, thus, those really semantically related queries cannot be grouped together effectively.

Query session is an extension to query and, accordingly, query clustering also is changed to "query session clustering". A query session is made up of the query and the subsequent activities the user performed. From accumulated web logs, a large amount of query sessions can be extracted using some user identification and session identification mechanisms. The click-through information in query sessions provides very valuable evidences to deduce users' search intentions more precisely than solely relying on query content. Although some naive methods based on clickthroughs in query sessions can improve the clustering results significantly, a more effective way is to combine query clustering and document clustering through query sessions. There are two ways to achieve such a combination — unidirectional clustering and bidirectional clustering. Unidirectional clustering utilizes document clustering to enhance query clustering unilaterally. On the contrary, bidirectional clustering employs an iterative, agglomerative method to cluster queries and documents simultaneously.

Content-based and session-based query clustering represent two different points of view and have their own advantages and shortcomings. Therefore, some combined measures taking advantage of both criteria may be more appealing. How to choose proper clustering algorithm in the query clustering context is also discussed, and four principal rules guiding the choice decision are given. Finally, a case study on Encarta web site is introduced in detail. The characteristics of different clustering methods are demonstrated and evaluated.

References

[1] D. Beeferman and A. Berger, Agglomerative clustering of a search engine query log, *Proceedings of the 6th ACM SIGKDD International Conference on Knowledge Discovery and Data Mining* (2000) pp. 407-416.

[2] R. Cooley, B. Mobasher, and J. Srivastava, Data preparation for mining World Wide Web browsing patterns, *Journal of Knowledge and Information Systems* Vol.1 No.1 (1999).

[3] H. Cui, J.-R. Wen, J.-Y. Nie, and W.-Y. Ma, Probabilistic query expansion using query logs, *Proceedings of the Eleventh World Wide Web conference (WWW 2002)* (2002) pp. 325-332.

[4] E. De Lima and J. Pedersen, Phrases recognition and expansion for short, precision-biased queries based on a query log, *Proceedings of the 22nd Annual International ACM SIGIR Conference on Research and Development in Information Retrieval* (1999) pp. 145-152.

[5] R. C. Dubes and A. K. Jain, *Algorithms for Clustering Data*, (Prentice Hill, 1988).

[6] M. Ester, H. Kriegel, J. Sander, and X. Xu, A density-based algorithm for discovering clusters in large spatial databases with noise, *Proceedings of the 2nd International Conference on Knowledge Discovery and Data Mining* (1996) pp. 226-231.

[7] M. Ester, H. Kriegel, J. Sander, M. Wimmer, and X. Xu, Incremental clustering for mining in a data warehousing environment, *Proceedings of the 24th International Conference on Very Large Data Bases* (1998) pp. 323-333.

[8] L. Fitzpatrick. and M. Dent, Automatic feedback using past queries: social searching? *Proceedings of the 20th Annual International ACM SIGIR Conference on Research and Development in Information Retrieval* (1997) pp. 306-312.

[9] G.W. Furnas, T.K. Landauer, L.M. Gomez, and S.T. Dumais, The vocabulary problem in human-system communication, *CACM* Vol.30 No.11 (1987) pp. 964-971.

[10] D. Gusfield, *Algorithms on Strings, Trees, and Sequences: Computer Science and Computational Biology, Part III "Inexact Matching, Sequence Alignment, and Dynamic Programming"*, (Press of Cambridge University, 1997).

[11] M.H. Hansen and E. Shriver, Using navigation data to improve IR functions in the context of web search, *Proceedings of the 10th International Conference on Information and Knowledge Management (ACM CIKM 2001)*, (2001) pp. 135-142.

[12] C.-K. Huang, L.-F. Chien, and Y.-J. Oyang, Query-session-based term suggestion for interactive web search, *WWW10 Poster Proceedings* (2001).

[13] K. Kukich, Techniques for automatically correcting words in text, *ACM Computing Surveys* Vol.24 No.4 (1992) pp. 377-439.

[14] V.A. Kulyukin, K.J. Hammond, and R.D. Burke, Answering questions for an organization online, *Proceedings of AAAI'98* (1998) pp. 532-538.

[15] D.D. Lewis and W.B. Croft, Term clustering of syntactic phrases, *Proceedings of the 13th Annual International ACM SIGIR Conference on Research and Development in Information Retrieval* (1990) pp. 385-404.

[16] Z. Lu and K. McKinley, Partial collection replication versus caching for information retrieval systems, *Proceedings of the 23rd Annual International ACM SIGIR Conference on Research and Development in Information Retrieval* (2000) pp. 248-255.

[17] G.A. Miller (eds.), WordNet: an on-line Lexical Database, *International Journal of Lexicography* Vol.3 No.4 (1990).

[18] R. Ng and J. Han, Efficient and effective clustering method for spatial data mining, *Proceedings of the 20th International Conference on Very Large Data Bases* (1994) pp. 144-155.

[19] P. Pirolli, J. Pitkow, and R. Rao, Silk from a sow's ear: Extracting usable structures from the Web. *Proceedings of 1996 Conference on Human Factors in Computing Systems (CHI-96)* (1996).

[20] J. Pitkow, In search of reliable usage data on the WWW, *Proceedings of the Sixth World Wide Web conference (WWW6)* (1997) pp. 451-463.

[21] M. Porter, An algorithm for suffix stripping, *Program* Vol.14 No.3 (1980) pp. 130-137.

[22] J. Rocchio, Relevance feedback in information retrieval, in G. Salton (eds.) *The Smart Retrieval System — Experiments in Automatic Document Processing* (Prentice-Hall Englewood Cliffs, 1971) pp. 313-323.

[23] G. Salton and C. Buckley, Improving retrieval performance by relevance feedback, *Journal of the American Society for Information Science*, Vol.41 No.4 (1990) pp. 288-297.

[24] G. Salton and M.J. McGill, *Introduction to Modern Information Retrieval* (McGraw-Hill Book Company, 1983).

[25] R. Srihari and W. Li, Question answering supported by information extraction, *Proceedings of TREC8* (1999) pp. 75-85.

[26] C.J. van Rijsbergen, *Information Retrieval (Second Edition)* (Butterworths, London, 1979).

[27] E. Voorhees, N.K. Gupta, and B. Johnson-Laird, Learning collection fusion strategies, *Proceedings of the 18th Annual International ACM SIGIR Conference on Research and Development in Information Retrieval* (1995) pp. 172-179.

[28] J.-R. Wen, J.-Y. Nie, and H.-J. Zhang, Clustering user queries of a search engine, *Proceedings of the Tenth World Wide Web conference (WWW10)* (2001) pp. 162-168.

[29] J.-R. Wen, J.-Y. Nie, and H.-J. Zhang, Query Clustering Using User Logs, *ACM Transactions on Information Systems (ACM TOIS)* Vol.20 No.1 (2002) pp. 59-81.

[30] P. Willett, Recent trends in hierarchical document clustering: A critical review, *Information Processing and Management* Vol.24 No.5 (1988) pp. 577-597.

[31] J. Xu and W.B. Croft, Query expansion using local and global document analysis, *Proceedings of the 19th Annual International ACM SIGIR Conference on Research and Development in Information Retrieval* (1996) pp. 4-11.

[32] O. Zamir and O. Etzioni, Web document clustering: A feasibility demonstration, *Proceedings of the 21st Annual International ACM SIGIR Conference on Research and Development in Information Retrieval* (1998).

Clustering and Information Retrieval (pp. 227-259)
W. Wu, H. Xiong and S. Shekhar(Eds.)

Clustering Techniques for Large Database Cleansing

Sam Y. Sung
Department of Computer Science
National University of Singapore, Singapore, 119260
E-mail: ssung@comp.nus.edu.sg

Zhao Li
Department of Computer Science
National University of Singapore, Singapore, 119260
E-mail: lizhao@comp.nus.edu.sg

Tok W. Ling
Department of Computer Science
National University of Singapore, Singapore, 119260
E-mail: lingtw@comp.nus.edu.sg

Contents

1 Introduction

Data cleansing, also called *data cleaning* or *data scrubbing*, deals with detecting and removing errors and inconsistencies from data in order to improve the quality of data [35]. It is a common problem in environments where records contain erroneous in a single database (e.g., due to misspelling during data entry, missing information and other invalid data etc.), or where multiple databases must be combined (e.g., in data warehouses, federated database systems and global web-based information systems etc.).

1.1 Motivation

The amount of data handled by organizations has been increasing at an explosive rate. The data is very likely to be dirty because of misuse of abbreviations, data entry mistakes, duplicate records, missing values, spelling errors, outdated codes etc [27]. A list of common causes of dirty data is described in [32]. As the example shown in [25], in a normal client database, some clients may be represented by several records for various reasons: (1) incorrect or missing data values because of data entry errors, (2) inconsistent value naming conventions because of different entry formats and use of abbreviations such "ONE" vs '1', (3) incomplete information because data is not captured or available, (4) clients do not notify change of address, and (5) client mis-spell their names or give false address (incorrect information about themselves). As a result, several records may refer to the same real world entity while not being syntactically equivalent. In [43], errors in databases have been reported to be up 10% range and even higher in variety of applications.

Dirty data will distort the information obtained from it because of the "garbage in, garbage out" principle. For example, in data mining, dirty data will not be able to provide data miners with correct information. Yet it is difficult for managers to make logical and well-informed decisions based on information derived from dirty data. A typical example given in [30] is the prevalent practice in the mass mail market of buying and selling mailing lists. Such practice leads to inaccurate or inconsistent data. One inconsistency is the multiple representations of the same individual household in the combined mailing list. In the mass mailing market, this leads to expensive and wasteful multiple mailings to the same household. Therefore, data cleansing is not an option but a strict requirement for improving the data quality and providing correct information.

In [21], data cleansing is identified as critical importance for many industries over a wide variety of applications, including marketing communications, commercial householding, customer matching, merging information systems, medical records etc. It is often studied in association with data warehousing, data mining and database integration. Especially, data warehousing [4, 20] requires and provides extensive support for data cleansing. They load and continuously refresh huge amounts of data from a variety of sources so the probability that some of the sources contain "dirty data" is high. Furthermore, data warehouses are used for decision making, so that the correctness of their data is vital to avoid wrong conclusions. For instance, duplicated or missing information will produce incorrect or misleading statistics. Due to the wide range of possible data inconsistencies, data cleaning is considered to be one of the biggest problems in data warehousing. In [38], data cleansing is identified as one of the database research opportunities for data warehousing into the 21st century.

1.2 Problem

Data cleansing generally includes many tasks because the errors in databases are wide and unknown in advance. One such main and most important task is to de-duplicate records, which recently receives much attention and many research efforts are focused on it. Before the de-duplication, there is a pre-processing stage which detects and removes any anomalies in the data records and then provide the most consistent data for the de-duplication. The pre-processing usually (but not limit to) does spelling correction, data type check, format standardization and abbreviation standardization etc. More discussions on the pre-processing are shown in Section 2. The discussion here is focused on the de-duplication.

Given the database having a set of records, the de-duplication is to detect all duplicates of each record. The duplicates include exact duplicates and also inexact duplicates. The inexact duplicates [31] are records that refer to the same real-world entity while not being synthetically equivalent. If consider the transitive closure, the de-duplication is to detect all clusters of duplicates and each cluster includes a set of records that represent the same entity. The computing of transitive closure is an option in some data cleaning methods, but an inherent requirement in some other data cleansing methods.

Formally, this de-duplication problem can be identified as follows. Let $\mathcal{D} = \{A_1, A_2, \cdots, A_N\}$ be the database, where A_i, $1 \leq i \leq N$, are records. Let $<A_i, A_j> = T$ denote that records A_i and A_j are duplicate, and

$$Dup(\mathcal{D}) = \{<A_i, A_j> \mid <A_i, A_j> = T, 1 \leq i, j \leq N \text{ and } i \neq j\}.$$

That is, $Dup(\mathcal{D})$ is the set of all duplicate pairs in $\mathcal{D}$. Then, given $\mathcal{D}$, the problem is to find out the $Dup(\mathcal{D})$.

Let $A_i \sim A_j$ be the equivalent relation among records that A_j is a duplicate record of A_i under transitive closure. That is $A_i \sim A_j$ iff there are records A_{i_1}, A_{i_2}, $\cdots$, A_{i_k}, such that $<A_i, A_{i_1}> = T$, $<A_{i_1}, A_{i_2}> = T$, $\cdots$, and $<A_{i_k}, A_j> = T$. Let $X_{A_i} = \{A_j | A_i \sim A_j\}$. Then $\{X_{A_i}\}$ are equivalent classes under this equivalent relation. Thus for any two records A_i and A_j, we have either $X_{A_i} = X_{A_j}$ or $X_{A_i} \cap X_{A_j} = \emptyset$. If the transitive closure is taken into consideration, the problem is then to find $TC(\mathcal{D}) = \{X_{A_i}\}$. More strictly, it is to find $TC_2(\mathcal{D}) = \{X_{A_i} || X_{A_i}| \geq 2\}$.

Clustering is a technique that groups a set of records such that the intracluster similarity is maximized and the intercluster similarity is minimized [5]. The data cleansing problem is then a clustering problem, which is to find all the clusters such that all records in the same cluster represent the same entity. However, the data cleansing problem has some special features that makes it a special clustering problem. From observing real world scenarios, each cluster normally only has a few records, and moreover, a lot of clusters are just singleton sets. That is, for large databases, $|TC(\mathcal{D})|$ is also large. One fundamental issue in data cleansing is that the size of the datasets involved may be very large and only a relatively small portion of the total available data can reside in main memory at any point in time. Therefore, all the representative records of clusters (one for each cluster) cannot reside in main memory at once. Furthermore, comparing each new record with all the representative records is very time-consuming. Thus normal clustering algorithms would not work for large database if considering the time taken.

1.3 Solution

Given a database, to detect exact duplicates is a simple process and is well addressed in [3]. The standard method is to sort the database and then check if the neighboring records are identical. The more complex process is to detect the inexact duplicates, which leads to two problems: (1) which records need to be compared and (2) how to compare the records to determine whether they are duplicate.

Thus, the (inexact) de-duplication consists of two main components: detection method and comparison method. A detection method determines which records will be compared, and a comparison method decides whether two records compared are duplicate.

In detection methods, the most reliable way is to compare every record with every other record. Obviously this method guarantees that all potential duplicate records are compared and then provides the best accuracy. However, the time complexity of this method is quadratic. It takes $N(N-1)/2$ comparisons if the database has N records, which will take very long time to execute when N is large. Thus it is only suitable for small databases and is definitely impracticable and infeasible for large databases. The clustering method that compares each record with the representative of each cluster is also impracticable for large database as we have explained in last section.

Therefore, for large databases, approximate detection algorithms that take far less comparisons (e.g., $O(N)$ comparisons) are required. Some approximate methods have been proposed [16, 17, 26, 30, 31]. All these methods have a common feature as they compare each record with only a limited number of records with a good expected probability that most duplicate records will be detected. All these methods can be viewed as the variances of "sorting and then merging within a window". The sorting is to bring potential duplicate records close together. The merging is to limit that each record is only compared with a few neighborhood records.

Based on this idea, Sorted Neighborhood Method (SNM) is proposed in [17]. SNM takes only $O(\omega N)$ comparisons by sorting the database on a key and making pair-wise comparisons of nearby records by sliding a window, which has size ω, over the sorted database. Other methods, such as Clustering SNM [17], Multi-pass SNM [17], DE-SNM [16] and Priority Queue [31] etc., are further proposed to improve SNM on different aspects (either accuracy or time). More discussions and analyses on these detection methods will be shown in Section 3.1.

The detection methods determine which records need to be compared, and pare-wise comparison methods are to decide whether two records compared are duplicated.

As shown in [17], the comparison of records to determine their equivalence is a complex inferential process that needs to consider much more

Name	Dept.	Age	Gender	GPA
Li Zhao	Computer Science	-	-	-
Li Zhai	Computer Science	-	-	-

Table 1: Two records with a few information known.

Name	Dept.	Age	Gender	GPA
Li Zhao	Computer Science	28	M	4.5
Li Zhai	Computer Science	28	M	4.5

Table 2: Two records with more information known.

information in the compared records than the keys used for sorting. The more information there is in the records, the better inferences can be made.

For example, for the two records in Table 1, the values in the "Name" field are nearly identical, the values in the "Dept." field are exactly the same, and the values in the other fields ("Age", "Gender" and "GPA") are unknown. We could either assume these two records represent the same person with a type error in the name of one record, or they represent different persons with similar name. Without any further information, we may perhaps assume the later. However, as the two records shown in Table 2, with the values in the "Age", "Gender" and "GPA" fields are known, we can determine that they represent the same person.

With the complex to compare records, one natural approach is using production rules based on domain-specific knowledge. Equational Theory was proposed in [17] to compare records. This approach uses a declarative rule language to specify an equational theory. Java Expert System Shell (JESS) [10], a rule engine and scripting environment, is employed by IntelliClean [24]. The rules are represented as declarative rules in the JESS engine. An example is given in Section 3.2.1

An alternative approach is to compute the degree of similarity for records. A *similarity function* $Sim : \mathcal{D} \times \mathcal{D} \mapsto [0, 1]$ is a function and satisfies

1. *reflexivity:* $Sim(A_i, A_i) = 1.0, \forall A_i \in \mathcal{D}$;

2. *symmetry:* $Sim(A_i, A_j) = Sim(A_j, A_i), \forall A_i, A_j \in \mathcal{D}$.

Thus the similar of records is viewed as the degree of similarity, which is a value between 0.0 and 1.0. Commonly, 0.0 means certain non-equivalence and 1.0 means certain equivalence [30]. A similarity function is *well-defined* if it satisfies 1) similar records will have large value (similarity) and 2) dissimilar records will have small value.

To determine whether two records are duplicate, a comparison method will typically just compare their similarity to a threshold, say 0.8. If their similarity is larger than the threshold, then they are treated as duplicate. Otherwise, they are treated as non-duplicate. Notice that the threshold are not given at random. It highly depends on the domain and the particular comparison methods in use.

Notice that the definition of Sim is abstractive and works for databases of any kind of data type. However, this approach is generally based on the assumption that the value of each field is a string. Naturally this assumption is true for a wide range of databases, including those with numerical fields represented in decimal notation. In [31], this assumption is also identified as a main domain-independent factor. Further note that rule-based approach can be applied on various data types, but currently, their discussions and implementations are only on string data as well since the string data is ubiquitous.

With this assumption, comparing two records is equal to compare two sets of strings where each string is for a field. Then any approximate string matching algorithms can be used as the comparison method.

Edit Distance [42] is a classic method in comparing two strings and has received much attention and widely used in many applications. Edit Distance returns an integer but the integer can be easily transfered (normalized) to a similarity value. The Smith-Waterman algorithm [39], a variant of edit distance, was employed in [31]. *Record Similarity (RS)* was introduced in [26], in which record equivalence is determined by viewing records similarity at three levels: token, field and record. The string value in each field is parsed as tokens by using a set of delimiters such as space and punctuations. Field weightage was introduced on each field to reflect the different importance. In Section 3.2, we will discuss these comparison methods in more details.

One issue should be addressed is that whether two records are equivalent is a semantical problem, i.e., whether they represent the same real-world entity. However, the record comparison methods which solve this problem depend on the syntax of the records (inexact duplicate). Hence, the syntactic calculations performed by the algorithms are only approximates of the real semantic equivalence problem [30]. In such calculations, errors are possible to occur, that is, correct duplicate records compared may not be discovered and false positives may be introduced.

All feasible detection methods, as we have shown, are approximate. Since none of the detection methods can guarantee to detect all duplicate records, it is possible that two records are duplicate but will not be detected. Further, all comparison methods are also approximate, as shown above, and none of them is completely trustworthy. Thus, no data cleansing method (consisting of detection methods and comparison methods) guarantees that it can find out exactly all the duplicate pairs, $Dup(\mathcal{D})$. It may not find some correct duplicate pairs and also introduce some false positives.

Given a data cleansing method, let $DR(\mathcal{D})$ be the duplicate pairs found by it, then $DR(\mathcal{D}) \cap Dup(\mathcal{D})$ is the set of correct duplicate pairs and $DR(\mathcal{D}) - Dup(\mathcal{D})$ is the set of false positives. Let $p_c = \frac{|DR(\mathcal{D}) \cap Dup(\mathcal{D})|}{|Dup(\mathcal{D})|}$ and $p_f = \frac{|DR(\mathcal{D}) - Dup(\mathcal{D})|}{|Dup(\mathcal{D})|}$. That is, the p_c and p_f are the percentages of correctness and falseness respectively. The p_c and p_f are two important parameters to determine whether a method is good enough, and whether a method is superior to another one. In addition, time is another important parameter and must be taken into consideration. Surely, comparing each record with every other record and using the most complicate rules as the data cleansing method will obtain the best accuracy. However, it is infeasible for large database since it cannot finish in reasonable time. Generally, more records compared and a more complicate comparison method used will obtain a more accuracy result, but this takes more time. Therefore, there is a tradeoff between accuracy and time.

All the works we introduced above are on the low-level of data cleansing, namely algorithm-level. Other work related to data cleansing includes proposing high level languages to express data transformation [6, 11, 14, 23], and introducing high level data cleansing frameworks [11, 24, 36]. The algorithms are fundamental in all data cleansing. For instance, in [24], multi-pass SNM is employed as its underlying detection system. In [11] edit distance is employed as its matching operator, and length filter [13] and multi-pass SNM are used as its matching operator optimization.

In the remainder of this article we will review what has been achieved in the field of data cleansing algorithms. In Section 2 we first simply discuss the pre-processing stage that is applied before cleansing. In Section 3 we then review the cleansing algorithms, including detection algorithms and comparison methods, in details. Finally, we conclude in Section 4.

2 Pre-processing

Given a database, before the de-duplication, there is generally a pre-processing on the records in the database. Pre-processing the records will increase the chance of finding duplicate records in the later cleansing. The pre-processing itself is quite important in improving the data quality. In [24], the pre-processing is identified as the first stage in the IntelliClean data cleansing framework.

The main task of the pre-processing is to provide the most consistent data for subsequent cleansing process. At this stage, the data records are first conditioned and scrubbed of any anomalies that can be detected and corrected. The techniques for the pre-processing are generally domain dependent. The following list shows the most common jobs and the relative techniques that can be performed in the pre-processing stage. Sometimes, some domain-specific jobs are required, which are different from database to database.

Spelling Correction: Some misspellings may exist in the database, such as "Singapore" may be mistakenly typed as "Singpore". Spelling correction algorithms have received a large amount of attentions for decades [2, 22]. Most of the spelling correction algorithms use a corpus of correctly spelled words from which the correct spelling is selected. The algorithm described by Bickel in [2] can be chosen for its simplicity and speed.

Data Type Check and Format Standardization: Data type check and format standardization can also be performed, such as, in the "data" field, 1 Jan 2002, 01/01/2002, and 1st January 2002 can be standardized to one fixed format. For this task, user may define a unique format for each field and all records are then processed and changed to it. For example, if "dd/mm/yyyy" is set as the format for the "data" field, 1 Jan 2002 and 1st January 2002 in records are then changed to 01/01/2002.

Abnormal Value Check: Abnormal values for fields can also be checked. For example, the range of age for human being is normally between 0 and 150. If there are records that have the age value beyond this, the value can be an error. Notice that the value ranges of the fields are domain dependent, and domain-specific knowledge and technique is then required.

Abbreviation	Word
NUS	National University of Singapore
CS	Computer Science
RD.	Road
RD	Road

Table 3: Example of an abbreviation file.

Inconsistent Abbreviation Standardization: The inconsistent abbreviations used in the data can also be resolved. For example, all occurrences of "Rd." and "Rd" in the address field will be replaced by "Road". Occurrences of 'M' in the "gender" field will be replaced by 'Male', and occurrences of 'F' will be replaced by 'Female'. An external source file containing the abbreviations of words is needed. Table 3 shows one example.

3 Data Cleansing

After the pre-processing, the database is further cleansed by de-duplicating records. The de-duplication consists of two components: the detection methods and comparison methods. The detection methods detect which records need to be compared and then let the comparison methods do the actual comparisons to determine whether the records are duplicate. Currently, the detection methods and the comparison methods are independent, that is, any detection method can be combined with any comparison method. With this independence, we can separate the discussions of the detection methods and comparison methods.

3.1 Detection Methods

For each record, only a very limited number of records compared with it are duplicate. As we have explained in Section 1.3, all existing (feasible) detection methods are approximate methods and they are the variances of "sorting and then merging within a window". However, they differ on deciding which records are needed to be compared.

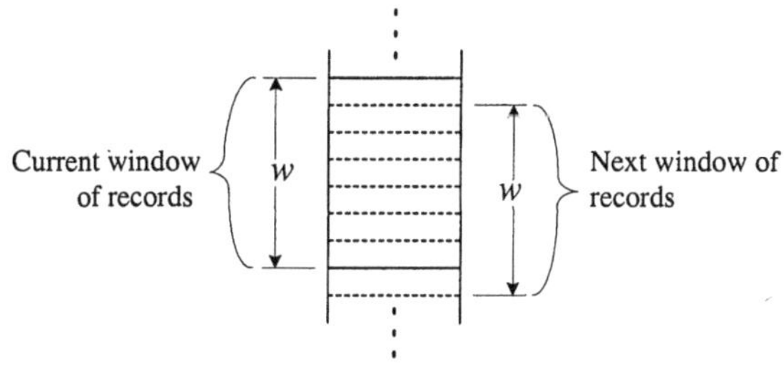

Figure 1: The merge phase of SNM.

Sorted Neighborhood Method

The *Sorted Neighborhood Method (SNM)* is proposed in [17]. One obvious method for bringing duplicate records close together is sorting the records over the most important discrimination key attribute of the data. After the sort, the comparison of records is then restricted to a small neighborhood within the sorted list. Sorting and then merging within a window is the essential approach of a Sort Merge Band Join as described by DeWitt [8]. In [17], SNM is summarized in three phases:

- Create Key: Compute a key for each record in the dataset by extracting relevant fields or portions of fields;
- Sort Data: Sort the records using the key;
- Merge: Move a fixed size window through the sorted list of records and limit the comparisons for duplicate records to those records in the window. If the size of the window is ω records, then every new record entering the window is compared with the previous $\omega - 1$ records to find duplicate records. The first record in the window slides out of the window (see Figure 1).

The effectiveness of this approach is based on the quality of the keys used in the sort. The key creation in SNM is a highly knowledge-intensive and domain-specific process [18]. Poorly chosen keys will result in a poor quality result, i.e., records that are duplicate will be far apart each other after the sort and then will not be discovered. For example, if the "gender" field in a database is chosen as the key, obviously, a lot of duplicate records would not be close together. Hence, keys should be chosen such that duplicate records have nearly equal key values. However, since the data to be cleansed is possibly corrupted, then the key extracted from the data will possibly be corrupted as well. Thus, a substantial number of duplicate records may not be detected.

Further, the "window size" used in SNM is an important parameter that affects the performance. Increasing the window size will increase the number of duplicate pairs found but also, on the other hand, increase the time taken. The performance result in [17] shows that the accuracy increases slowly but the time increases fast when increasing the window size. Thus, increasing the window size does not help much if taking into consideration that the time complexity of the procedure goes up fast as the window size increase, and it is fruitless at some point to use a large window.

Clustering SNM

As the database becomes very large, sorting the data may take a great amount of time although it may not be the dominant cost of cleansing. In [17], the authors considered an alternative to sorting based upon first partitioning the dataset into independent clusters using a key extracted from the data. Then SNM is applied to each individual cluster independently. This method is called as *Clustering SNM* and can be summarized as the following two phases:

1. **Cluster Data:** Scan the records in sequence and extract a key for each record and then using the extracted key to partition the records into independent subsets of data (clusters).

2. **SNM:** Apply SNM independently on each cluster. The key does not need to be recomputed, and the key extracted above can be used for sorting (see Figure 2).

Notice that the "Cluster Data" phase is simply to partition the data into a set of independent clusters and any existing clustering methods can be used. For example, an approach that closely resembles the multidimensional partitioning strategy of [12] can be used. Or a hash method on the key can also be used. However, one issue should be considered is that how well the partitioned data is after clustering. The clusters should be partitioned such that as much as possible duplicate records appear in the same cluster. So the clustering method used need to be carefully chosen and performance tests

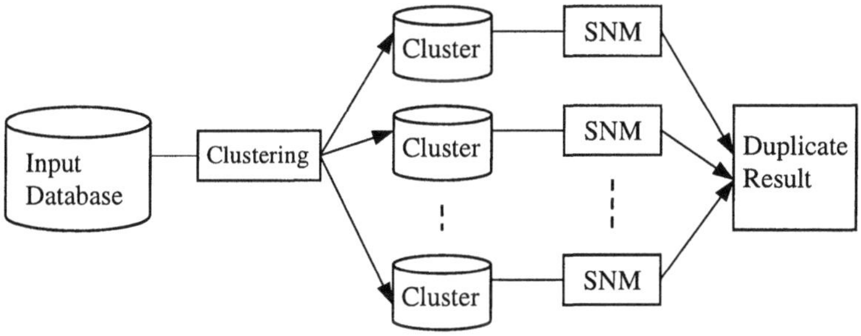

Figure 2: Clustering SNM.

may also needed. In addition, that how many clusters should be partitioned depends on the size of the database and also the size of the main memory. Normally, the size of each cluster partitioned should be smaller than that of main memory. Then the sorting on each cluster can be a main-memory based process.

Since the dataset is partitioned into small clusters and do not need a completely sorted database, the clustering SNM takes less time than SNM (sorting some small datasets is faster than sorting a large dataset). Furthermore, dataset is partitioned into a set of small clusters, which is ideally for parallelization. However, two duplicate records may be partitioned into two different clusters, then they cannot be detected, which results in a decrease of the number of correct duplicate results. Thus the clustering SNM provides the trade-off between time and accuracy.

Multi-pass SNM

Generally, no single key will be sufficient to catch all duplicate records and the number of duplicate records missed by one run of the SNM can be large [17]. For instance, if a customer has two records in a customer database, one with "FIN NO" 157293578 and another with "FIN NO" 517293578, and if the "FIN NO" is chosen as the key, then both records are very likely not in the same window, i.e., these two duplicated records will not be detected.

To increase the number of duplicate records detected, *Multi-pass SNM* is then proposed in [17]. Multi-pass SNM is to execute several independent runs of SNM, each using a different key and a relatively small window. Each independent run will produce a set of pairs of duplicate records. The results is the union of all pairs discovered by all independent runs, plus all those pairs that can be inferred by transitive closure. The transitive closure is executed on pairs of record id's, and fast solutions to compute transitive closure exist [1, 31].

This approach works based on the nature of errors in the data. One key having errors will lead to that some duplicate records cannot be detected. However, in such records, the probability of error appearing in another key of the records may not be so large. Therefore, the duplicate records missed in one pass would be detected in another pass with different key. So multi-pass increases the correctness. As the example shown above, if the name in the two records are the same, then a second run with the name field as the key will detect them correctly as duplicate records. Theoretically, suppose the probability of duplicate records missed in one pass is p_ω, $0 \leq p_\omega \leq 1$, where ω is the window size, then the probability of duplicate records missed in n independent passes is p_ω^n. So, the correctness for n-passes is $1-p_\omega^n$, while the correctness for one pass is $1-p_\omega$. Surely, $1-p_\omega^n$ is larger than $1-p_\omega$. For example, if $n=3$ and $p_\omega = 50\%$, we have $1-p_\omega^n = 1-0.5^3 = 87.5\%$ and $1-p_\omega = 1-0.5 = 50\%$.

The performance result in [17] shows that multi-pass SNM can drastically improve the accuracy of the results of only one run of SNM with varying large windows. Multi-pass SNM can achieve p_c higher than 90%, while SNM generally only gets p_c about 50% to 70%. Particularly, only a small window size is needed for the multi-pass SNM to obtain high accuracy, while no single run with a key can produce comparable accuracy results with a large window.

One issue in Multi-pass SNM is that it employs transitive closure to increase the number of duplicate records. The transitive closure allows duplicate records to be detected even without being in the same window during an individual window scan. However, the duplicate results obtained may contain errors (false positives), as explained in Section 1.3 that no comparison methods are completely trustworthy, and transitive closure propagates the errors in results. Thus, multi-pass SNM also increases the number of false positives.

Duplication Elimination SNM

Duplicate Elimination SNM (DE-SNM) [16] improves SNM by first sorting the records on a chosen key and then dividing the sorted records into two

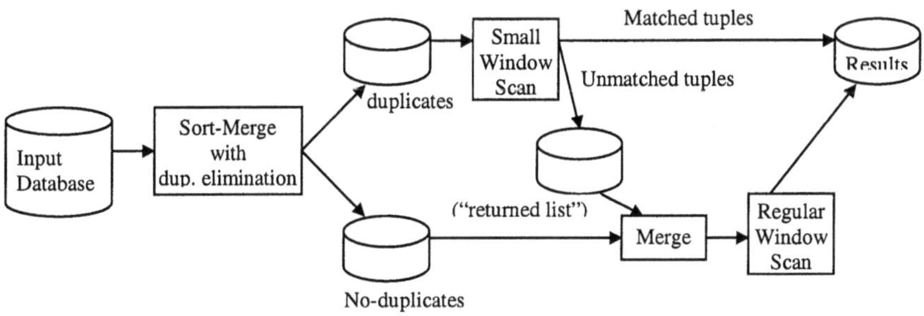

Figure 3: Duplication Elimination SNM.

lists: a duplicate list and a non-duplicate list. The duplicate list contains all records with exact duplicate keys. All the other records are put into the non-duplicate list. A small window scan is first performed on the duplicate list to find the lists of matched and unmatched records. The list of unmatched records is merged with the original non-duplicate list and a second window scan is performed. Figure 3 shows how DE-SNM works.

DE-SNM does not contribute much on the improvement of accuracy of SNM. The benefit of DE-SNM is on that it runs faster than SNM under the same window size, especially for the databases that are heavily dirty. If the number of records in duplicate list is large, DE-SNM will run faster than SNM.

Priority Queue Method

Under the assumption of transitivity, the problem of detecting duplicates in a database can be described in terms of determining the connected components of an undirected graph. Transitivity of the "is a duplicate of" relation is equivalent to reachability in the graph. There is a well-known data structure, *union-find data structure* [7, 40], that efficiently solves the problem of determining and maintaining the connected components of undirected graph. This data structure keeps a collection of disjoint updatable sets, where each set is identified by a representative member of the set. The data structure has two operations *Union(x,y)* and *Find(x)*. More information on the union-find data structure can be found in [7].

By using the union-find data structure, *Priority Queue method* is suggested in [31]. Priority Queue does two passes of sorting and scanning. Two passes are used to increase the accuracy over one pass as the reason is shown in multi-pass SNM. The first pass treats each record as one long string and sorts the string lexicographically, reading from left to right. The second pass does the same reading but from right to left. Unlike previous algorithms, the sorting of the records in each pass is domain-independent. Thus the Priority Queue is a domain-independent detection method.

Priority Queue scans the database sequentially and determines whether each record scanned is or is not a member of a cluster represented in a priority queue. To determine cluster membership, it uses the *Find* operation. If the record is already a member of a cluster in the priority queue, then the next record is scanned. If the record is not already a member of any cluster kept in the priority queue, then the record is compared to representative records in the priority queue using the Smith-Waterman algorithm [39]. If one of these comparisons succeeds, then the record belongs in this cluster and the *Union* operation is performed on the two sets. On the other hand, if all comparisons fail, then the record must be a member of a new cluster not currently represented in the priority queue. Thus the record is saved in the priority queue as a singleton set. For practical reasons, the priority queue contains only a few number (e.g. 4) of sets of records (like the window size in SNM), and the sets in the priority queue represent the last few clusters detected.

Priority Queue using the union-find data structure to compute the transitive closure online, which may result in saving a lot of unnecessary comparisons. For example, for three duplicate records A_1, A_2 and A_3, there are three comparisons in SNM. However, in Priority Queue, if A_1 and A_2 have been compared and *Union*ed in a cluster, in which A_1 is the representative, then when A_3 is scanned, it only needs to compare with A_1 and one comparison is saved. Note that if the database is clean or slightly dirty, then each cluster in the priority queue most likely contains only one record (singleton set). Under this conditions, the Priority Queue is just the same as the Multi-pass SNM (2 passes) but with extra cost on the *Union* and *Find* operations. Thus for clean or slightly dirty databases, Priority Queue does not provide any help, or even worse because it takes more time for the extra *Union* and *Find* operations before each comparison. However, surely, Priority Queue works better for heavily databases since clusters likely contain more than one record.

In Priority Queue, the size of the priority queue should be determined. Thus it still faces the same "window size" problem as SNM does. Further, as Priority Queue computes transitive closure online, it faces the transitive

Condition and requirement	Suggestion
The database is quite small, or it is large but long execution time is acceptable	Pair-wise comparisons
The database is very large, less false positives are more important than more correctness, and multiple processors are available	Clustering SNM
More correctness would be better, and some false positives are acceptable	Multi-pass SNM
The database is heavily dirty, and some false positives are acceptable	Priority Queue

Table 4: The methods would be used for different conditions.

closure problem (discussed in Multi-pass SNM) as well. Moreover, representative records are chosen for each cluster and heuristics need to be developed for choosing the representative records, which will affect the results greatly.

3.1.1 Summary

We have introduced the detection methods and shown that each has its own tradeoff. Due to that the pair-wise comparison of every record with every other record is infeasible for large databases, SNM is firstly proposed by providing an approximate solution. SNM includes three phases: Create Key, Sort Data, and Merge. The "Sorting" performs the first clustering on the database such that the similar records are close together. Then the "merging" performs clustering again on the sorted database to obtain the clustering result such that the records in each cluster represent the same entity and the records in different clusters represent different entities. The sorting and merging together is two-level clustering that the sorting is the first loose clustering, while the merging is the second strict clustering. In sorting, only the key value (normally one field) need to be compared, while in merging, all fields should be considered.

Clusterings (sorting and merging) are used to significantly reduce the detection scope and comparison time with achieving a reasonable accuracy. SNM generally cannot obtain high accuracy but works for any database coherently. Other approximate methods are further proposed to improve the performance on either efficiency or accuracy. Multi-pass SNM can largely increase the accuracy under the same time than SNM does. Priority Queue can save some unnecessary comparisons taken by SNM by computing the transitive closure online. Priority Queue may be faster than SNM but cannot improve the accuracy under the same conditions with SNM. In addition, the performance of Priority Queue depends on the degree of dirty of databases. For clean and slightly dirty databases, Priority Queue does not provide any help. But for dirty databases, Priority Queue is much faster. The more dirty the database is, the more time it can save. Like Priority Queue, DE-SNM can also run faster than SNM for dirty databases, but DE-SNM will decrease the accuracy. Clustering SNM is an alternative method. As the name shows, Clustering SNM does one even looser clustering before applying SNM. The clustering SNM does three level clustering from looser to stricter. Clustering SNM is faster than SNM for very large databases but it may decrease the accuracy as well. Further, Clustering SNM is suitable for parallel implementation.

Given the trade-off of each method, a natural question is, under certain conditions, which method should be employed. Table 4 gives some suggestions. Practically, among all these methods, multi-pass SNM is the most popular one. Some data cleaner systems, such as IntelliClean [24], DataCleanser DataBlade Module [29] etc., employ it as their underlying detection system.

3.2 Comparison Methods

As the detection methods determine which records need to be compared, pare-wise comparison methods are then used to decide whether two records compared are duplicate. As we have indicated in Section 1.3 that the comparison methods can be distinguished as two different approaches, namely rule-based and similarity-based. The rule-based approach is using production rules based on domain-specific knowledge, and the similarity-based approach is by computing the degree of similarity of records, which is a value between 0.0 and 1.0.

Notice that the comparison of records is quite complicated, it needs to take more information into consideration than the sorting does in the detection methods. Thus, the cost of comparisons is the dominate of the time taken by cleansing, which is proven by the performance studies in [17]. This further shows that the importance on avoiding unnecessary calls to the record comparison function by the detection system.

Further, all comparison methods (either rule-based or similarity-based) are only approximate methods. That is, none of them can guarantee to discover exactly correct result, which means that, given two duplicate records, the comparison method may not detect them as duplicate, or given two non-duplicate records, the comparison method may detect them as duplicate. The reason is that whether two records are duplicate is a semantical problem, but the solution to it is syntactical based.

3.2.1 Rule-based Methods

The rule-based approach uses a declarative rule language to specify the rules. A rule is generally of the form:

```
if <condition>
then <action>
```

The action part of the rule will be activated when the conditions are satisfied. Complex predicates and external function references may be contained in both the condition and action parts of the rule. The rules are derived naturally from the business domain. The business analyst with the subject knowledge is able to fully understand the governing business logic and can develop the appropriate conditions and actions.

Equational Theory was proposed in [17] to compare records. The following is a simplified rule[17] that describes one axiom of equational theory:

```
Given two records, r1 and r2.
IF the last name of r1 equals the last name of r2,
   AND the first names differ slightly,
   AND the address of r1 equals the address of r2
THEN
   r1 is equivalent to r2.
```

The implementation of "differ slightly" is based on the computation of a distance function applied to the first name fields of two records, and the comparison of its results to a threshold to capture obvious typographical errors that may occur in the data. The selection of a distance function and proper threshold is a knowledge intensive activity that demands experimental evaluation. An improperly chosen threshold will lead to either an increase in the number of false positives or to a decrease in the number of correct duplicate records.

In [17], rules are written in OPS5 [9]. In [25], rules are written in the Java Expert System Shell (JESS) [10]. JESS is a rule engine and scripting environment written in Sun's Java language and was inspired by the CLIPS [37] expert system shell. The data cleansing rules are represented as declarative rules in the JESS engine. The following shows one such rule (presented in [25] in pseudocode) written in JESS engine.

```
INPUT RECORDS: A, B
IF
    (A.currency == B.currency) AND
    (A.telephone == B.telephone AND
    (A.telephone != EMPTY_STRING) AND
    (SUBSTRING_ANY(A.code, B.code) == TRUE) AND
    (FIELDSIMILARITY(A.address, B.address) > 0.85)
THEN
    DUPLICATES(A, B)
```

For the rule to be activated, the corresponding currencies and telephone numbers must match. Telephone numbers must also not be empty, and one of the codes must be a substring of the other. The address must also be very similar.

The effectiveness of the rule-based comparison method is highly dependent on the rules developed. As well-developed rules are effective in identifying true duplicates and also strict enough to keep out false positives, the not well-developed rules will introduce even worse results. Therefore, the rules should be carefully developed and generally are tested repeatedly for the particular domain. As a result, the process of creating such (well-developed) rules can be time consuming. Further, the rules must be continually updated when new data is added to the database and it does not satisfy the previous defined rules. Moreover, the rule-based comparison methods are quite slow and do not clearly scale up for very large datasets. For example, in the experimental study in [17], all the rules are first written in OPS5 and then translated by hand into C since the OPS5 compiler is too slow.

To avoid these disadvantages in the rule-based approach, similarity approach is an alternative. Although the similarity-based methods can resolve the disadvantages in rule-based method, they have their own disadvantages that we will show later. In the following, we discuss and analyze the similarity-based methods in details.

Operations:	intention
delete i –	
	ntention
substitute n by e –	
	etention
substitute t by x –	
	exention
substitute n by c –	
	exection
insert u –	
	execution

Figure 4: The operations taken by transforming "intention" to "execution".

3.2.2 Similarity-based Methods

Similarity-based approach is to compute the degree of similarity for records by a similarity function Sim, defined in Section 1.3, which returns a value between 0.0 and 1.0. Two records having large Sim value means that they are very similar. In the special values, 0.0 means absolute non-equivalence and 1.0 means absolute equivalence. Notice that the definition of Sim can be applied on any data type, such as strings and images etc. Therefore, how to view the content of records is important for the definition of similarity function. Of course, the similarity function for two strings is definitely different with the similarity function for two images.

Due to the string data is ubiquitous, currently the discussion is focused on this type of data. In the following discussion, we can assume that each field only contains string value. Thus, record comparison is basically an string matching algorithm and any of the approximate string matching algorithms can be used as the record comparison method with the detection method.

Edit Distance

Edit Distance [42] is a classic method in comparing two strings that has received much attention and has applications in many fields. It can also be employed in data cleansing and is a useful measure for similarity of two strings.

```
int edit(char* x, char* y) \* computation of edit distance *\
{
   int m = strlen(x), n = strlen(y);
   int EDIT[m][n], delta;
   for (i = 1; i <= m; i++) EDIT[i, 0] = i;
   for (j = 1; j <= n; j++) EDIT[0, j] = j;
   for (i = 1; i <= m; i++) {
      for (j = 1; j <= n; j++) {
         if (x[i] == y[j]) delta = 0;
         else delta = 1;
         EDIT[i, j] = min(EDIT[i-1, j]+1, EDIT[i, j-1]+1,
                          EDIT[i-1, j-1]+delta);
      }
   }
   return EDIT[m, n]
}
```

Figure 5: The dynamic programming to compute edit distance.

Edit distance is defined as the minimum number of insertions, deletions, and substitutions needed to transform one string into another. For example, the edit distance between "intention" and "execution" is five. Figure 4 shows the operations taken by transforming "intention" to "execution". Edit distance is typically implemented using dynamic programming [15], and run in $O(mn)$ time where m and n are the lengths of the two strings. Figure 5 shows the dynamic programming to compute the Edit Distance.

For two strings, Edit Distance returns an integer value. However, the lengths of the strings compared need to be taken into account. Two strings of length 10 differing by 1 character have the same edit distance as two strings of length 2 differing by 1 character, but we would most likely state that only the length 10 strings are almost equal. Thus edit distance need to be normalized. *Post-normalization* by the maximal length of the compared strings is quite popular. Since the edit distance of any two strings is between 0 and the maximal length of them, the post-normalization returns a value between 0.0 and 1.0. A variant normalization method, called *normalized edit distance* (NED), is proposed in [28]. NED has its advantage but its computation requires significantly more work than that of post-normalized edit distance. The post-normalization returns a distance value between 0.0 and 1.0 (i.e., 0.0 means absolute equivalence), which can be easily transfered to similarity as the difference of one and the normalized value.

The post-normalized edit distance is to compare two strings instead of two sets of strings, thus it works on the field level instead of the record level (each field is a string, and each record is a set of strings). To make it work on the record level, in [31], each record is viewed just as one big string, and *Smith-Waterman algorithm* [39] is employed to compare records. The Smith-Waterman algorithm is a variant of edit distance and was originally developed for finding evolutionary relationships between biological protein and DNA sequences. Record Similarity, which is discussed as follows, uses another solution, which is to assign weightages for fields and then the similarity of records equals to the sum of the similarity of field times the field weightage. More information can be found later.

Record Similarity

Record Similarity (RS) was introduced in [26]. In Record Similarity, record equivalence can be determined by viewing records at three levels: token, field and record. Since each field has different importance, field weightage is introduced on all the fields. The field weightage is decided through experimental tests and the sum of all field weightages equals to 1. The string value in each field is parsed as tokens by using a set of delimiters such as space and punctuations. Tokens can be viewed as meaningful components. For example, suppose the delimiter is space, then the string "Li Zhao" has tokens {"Li", "Zhao"}.

The process of computing the similarity between two records begins with comparing the sorted tokens of the corresponding fields. The following shows the details in computing the three levels similarities.

(1) Compute Token Similarity:

- If two tokens t_1 and t_2 are exactly matched, then they have a degree of similarity of 1;
- Otherwise, if there is a total of x characters in the token t_1, then we deduct $1/x$ from the maximum degree of similarity ($DoS_{(t_1,t_2)}$) of 1 for each character that is not found in the other token t_2.

In this definition, the similarity of tokens is not symmetric. That is, $DoS_{(t_i,t_j)} \neq DoS_{(t_j,t_i)}$. For example, if two tokens "cat" and "late" are compared, then the degree of similarity of comparing "cat" with "late", $DoS_{(cat,late)}$ = 1-1/3 = 0.67 since the character c in "cat" is not found in "late", and $DoS_{(late,cat)}$ = 1-2/4 = 0.5 since the characters l and e are not found in "cat".

(2) Compute Field Similarity:

- Suppose a field F in record X has tokens $x_1, x_2, \cdots, x_n$, and the field F in record Y has tokens $y_1, y_2, \cdots, y_m$;
- Each token x_i, $1 \leq i \leq n$, is compared with all the tokens y_j, $1 \leq j \leq m$;
- Let DoS_{x_1} be the maximum of the degree of similarities computed for tokens x_1 with $y_1, y_2, \cdots, y_m$ respectively. That is, $DoS_{x_1} = max\{DoS_{(x_1,y_1)}, DoS_{(x_1,y_2)}, \cdots, DoS_{(x_1,y_m)}\}$. Similarly, we have $DoS_{x_2}, \cdots, DoS_{x_n}, DoS_{y_1}, DoS_{y_2}, \cdots, DoS_{y_m}$;
- Field similarity for records X and Y on this field F is given by $Sim_F(X,Y) = \frac{\sum_{i=1}^{n} DoS_{x_i} + \sum_{j=1}^{m} DoS_{y_j}}{n+m}$.

(3) Compute Record Similarity:

- Suppose the records have r fields, $F_1, F_2, \cdots, F_r$, and the field weightages are $W_1, W_2, \cdots, W_r$ respectively, where $\sum_{i=1}^{r} W_i = 1$;
- Record Similarity for records X and Y is given by $Sim(X,Y) = \sum_{i=1}^{r}(Sim_{F_i}(X,Y) \times W_i)$.

Although the token similarity is not symmetric, the field similarity and the record similarity are symmetric. That is, for any field F and two records X and Y, we have $Sim_F(X,Y) = Sim_F(Y,X)$ and $Sim(X,Y) = Sim(Y,X)$. Two records are treated as a duplicate pair if their record similarity exceeds a certain threshold such as 0.8.

Like Edit Distance, the time complexity for Record Similarity is also $O(mn)$ for two strings with length m and n. Since each field needs first to be partitioned into tokens, Record Similarity runs a little slower than Edit Distance.

Notice that the method for computing the similarity of records is extensible. The same method can be used for transferring any field level similarity to record level similarity. That is, for any field similarity method Sim_F, such as Edit Distance, it can be transfered to record similarity as $Sim = \sum_i(Sim_{F_i} \times W_i)$.

Record Similarity is employed in IntelliClean and generally shows good performance. However, theoretically, Record Similarity is not well defined.

Record	Field
X	ab ex
Y	ex ex ex ex ex ex ex ex ex ex

Table 5: Tokens repeat problem in Record Similarity.

It surely guarantees that similar records have large value. However, it may also assign some (very) dissimilar records with large value. Especially, it does not address the repeated tokens well. For example, the two records (for simplicity, suppose that they have only one field) in Table 5 are very dissimilar, but the Record Similarity computed for them is really large. Suppose the delimiter is space. For record X, the token similarities for "ab" and "ex" are $DoS_{ab} = 0$ and $DoS_{ex} = 1$ respectively. For record Y, there are 10 repeated tokens "ex" and each has $DoS_{ex} = 1$. Thus, we have $Sim(X,Y) = \frac{1+10}{2+10} = 0.92$, which is a quite large value, with which a false positive may be introduced.

Moving Contracting Window Patter Algorithm

The *Moving Contracting Window Patter Algorithm (MCWPA)* is proposed in [34]. The MCWPA is focused on the field level. Unlike Record Similarity which is a token base similarity method, MCWPA is a substring based method.

All characters as a whole within the window (string) constitute a *window pattern*. As an example, for the string "abcde", when the window is sliding from left to right with the window size being 3, the series of window patterns obtained are "abc", "bcd" and "cde". Notice that "window size" used here is different with that used in detection methods (e.g., SNM).

Given a Field F and two records X and Y, Let X_F denote the string value of field F of record X. Suppose that X_F and Y_F have m and n characters respectively (including blank space or comma). The field similarity for F of X and Y is given as

$$Sim_F(X,Y) = \frac{2\sqrt{SSNC}}{m+n}.$$

The algorithm to calculate $SSNC$ is shown in Figure 6. $SSNC$ represents the Sum of the Square of the Number of the same characters between X_F and Y_F. $Sim_F(X,Y)$ reflects the ratio of the total number of the common characters in two fields to the total number of characters in two fields.

```
1.  w = the smaller of n and m;
2.  SSNC = 0;
3.  s = the smaller of X_F and Y_F;
4.  window is placed on the leftmost position;
5.  while ((w != 0) and (still some characters in s are accessible))
6.  {
7.      while (window right border does not exceed the right border of s)
8.      {
9.          if (the window pattern in X_F has the same pattern anywhere in Y_F)
10.         {
11.             SSNC = SSNC + w^2;
12.             mark the pattern characters in X_F and Y_F as inaccessible
                characters to avoid revisiting;
13.         }
14.         move window rightward by 1 (if the window contains an inaccessible
            character, then keep moving window rightward until window contains
            only accessible characters)
15.     }
16.     w = w - 1;
17.     window is placed on the leftmost position where the window contains
        accessible characters;
18. }
19. return SSNC;
```

Figure 6: Calculate $SSNC$ in MCWPA algorithm.

MCWPA does not have the repeated tokens problem existed in Record Similarity. For the two records in Table 5, since they only have one common length-three substring " ex", then the similarity computed by MCWPA for them is $Sim_F^{MCWPA}(X,Y) = \frac{2\sqrt{3^2}}{29+29} = \frac{3}{29} = 0.1$, while the similarity computed by Records Similarity is $Sim_F^{RS}(X,Y) = 0.92$. So MCWPA is more accurate to describe the two records in Table 5 and will not obtain them as duplicate.

As MCWPA can achieve better accuracy than Record Similarity, its efficiency depends on the strings compared. It may take more time in the worst case and less time in the best case. For two strings with length m and n respectively and suppose $m \geq n$, the time complexity (best and worst) for Record Similarity is always $O(mn)$, while the worst time complexity for

MCWPA is $O(n^2m)$ and the best time complexity is $O(n)$. Generally, the more common substring the two strings have, the faster MCWPA is.

We have shown some similarity methods, Edit Distance, Record Similarity, and MCWPA. Other existed string comparison methods can also be used as the comparison method, such as Longest Common Subsequence [19] and Longest Common Substring.

A *subsequence* [33] of a given string is a string that can be obtained by deleting zero or more symbols from the given string. The *Longest Common Subsequence (LCS)* is to find the maximum length of a common subsequence of two strings, while the *Longest Common Substring* is to find the maximum length of a common substring of two strings.

Like Edit Distance, the Longest Common Subsequence is also solved by the dynamic programming technique. Thus its time complexity is $O(mn)$ as well for two strings with lengths m and n respectively. The Longest Common Substring problem can be solved in linear time using a data structure known as the suffix tree [41] but the solution is extremely complicated.

3.2.3 Summary

As we can see, the similarity approach is simply to compute a similarity value, and it has the following advantages:

- *Easy implementation:* It can be easily implemented embedded in any data cleansing system.

- *Uniform and stable:* For different databases or updated databases, only the field weightages and threshold need to be reset. Since the field weightages and threshold can be implemented as input parameters, the method itself does not need to be modified.

- *Fast and scalable:* It is much faster than rule-based approach and scales well for large databases.

However, it also has disadvantage. Since similarity methods only return a similarity value, they are hard to achieve the same accuracy as the well-developed rules does. For example, for two non-duplicate records that are very similar, the similarity methods will return a value that is larger than the given threshold. Thus the non-duplicate records are detected as duplicate, which is a false-positive.

In general, it is difficult to achieve both in terms of efficiency and accuracy. The rule-based approach and similarity-based approach show the trade-off between time and accuracy.

Among these similarity methods, each has its usage and it is hard to say which one is superior to another one. Given two records with values "Li Zhao" and "Zhao Li" in the "name" field respectively. Whether they are equivalence or not depends on the knowledge of the databases. They could be equivalence in some databases but be non-equivalence in some other databases. Record Similarity will treat them as equivalence, while Edit Distance more likely will treat them as non-equivalence. Thus Record Similarity may be suitable for some databases and Edit Distance is suitable for some others.

Further, Record Similarity could not address the repeated tokens in field well, while MCWPA does not have such problem. Thus MCWPA could improve the accuracy for that kind of databases having lots of repeated tokens in field. However, this kind of databases are not common in real case. In addition, Edit Distance does not have this problem as well.

4 Conclusion

In this article, we have reviewed some important data cleansing methods. The discussion is focused on the algorithm level, which is the fundamental for all data cleansing methods.

Data cleansing generally consists of many tasks due to the errors in database are extensive and unknown in advance. One such main task is to de-duplicate records. Generally, before the de-duplication, there is a pre-processing on the records, which detects and removes any anomalies in the data records. This pre-processing not only improves the data quality but also provides the most consistent database for the de-duplication, which can increase the chance of finding duplicate records.

The de-duplication is a (special) clustering problem. The result obtained contains a set of clusters such that records in the same cluster represent the same entity and records in the different clusters represent different entities. The de-duplication can be divided into two independent components, detection methods and comparison methods.

The need for detection methods is due to that the databases to be cleansed could be very large. For small databases, pair-wisely comparing every record with every other record can provide the best result, but it is infeasible for large database. Thus approximate detection methods are proposed. In Section 3.2, we have discussed some important existed detection methods in details, including SNM, Clustering SNM, Multi-pass SNM, DE-SNM and Priority Queue. All these methods are based on the basic idea "sorting and then merging within a window" and each method has its own trade-off between accuracy and time. As we have explained, the "sorting and then merging within a window" is a two-level clustering method, while "sorting" stands for a looser clustering and "merging" is a further stricter clustering based on the looser clustering. The two-level clustering can largely decrease the execution time while still guarantees to obtain an acceptable result.

Like other clustering methods in Information Retrieval, a comparison method (measure method) is needed in data cleansing as well. When the records needed to be compared, the comparison methods are then to do the actual comparison. Since whether two records are duplicate is a semantical problem and the solutions are syntactical based, all comparison methods are only approximate and none of them is completely trustworthy. The comparison methods can be divided into two different approaches, rule-based and similarity-based. Each approach has its own advantages and disadvantages. Similar to the detection methods, the rule-based approach and similarity-based approach show the trade-off between time and accuracy as well.

With the clustering method by combining the detection methods and comparison methods, data can be efficiently and effectively cleansed, and then the best data quality is provided.

References

[1] R. Agrawal and H. V. Jagadish. Multiprocessor transitive closure algorithms. In *Proc. Int'l Symp. On Databases in Parallel and Distributed Systems*, pages 56–66, December 1988.

[2] M. A. Bickel. Automatic correction to misspelled names: a fourth-generation language approach. *Communications of the ACM*, pages 30(3):224–228, 1987.

[3] D. Bitton and D. J. DeWitt. Duplicate record elimination in large data files. *ACM Transactions on Database Systems*, pages 8(2):255–265, 1983.

[4] S. Chaudhuri and U. Dayal. An overview of data warehousing and olap technology. In *ACM SIGMOD Record*, page 26 (1), 1997.

[5] M. S. Chen, J. Han, and P. S. Yu. Data mining: An overview from database perspective. *IEEE Transactions on Knowledge and Data Engineering*, pages 8(6):866–883, 1996.

[6] W. Cohen. Integration of heterogeneous databases without common domains using queries based on textual similarity. In *Proceedings of the ACM SIGMOD International Conference on Managemnet of Data*, 1998.

[7] T. H. Cormen, C. E. Leiserson, and R. L. Rivest. *Introduction to Algorithms.* MIT Press, 1990.

[8] D. J. DeWitt, J. F. Naughton, and D. A. Schneider. An evaluation of non-equijoin algorithms. In *Proc. 17th Int'l. Conf. on Very Large Databases*, pages 443–452, Barcelona, Spain, December 1991.

[9] C. L. Forgy. Ops5 user's manual. Technical Report CMU-CS-81-135, Carnegie Mellon University, July 1981.

[10] E. J. Friedman-Hill. Jess, the java expert system shell, 1999. Available from `http://herzberg.ca.sandia.gov/jess`.

[11] H. Galhardas, D. Florescu, D. Shasha, E. Simon, and C. A. Saita. Declarative data cleaning: Language, mode, and algorithms. In *Proc. 27th Int'l. Conf. on Very Large Databases*, pages 371–380, Roma, Italy, 2001.

[12] S. Ghandeharizadeh90. *Physical Database Design in Multiprocessor Database Systems.* PhD thesis, Department of Computer Science, University of Wisconsin - Madison, 1990.

[13] L. Gravano, P. G. Ipeirotis, H. V. Jagadish, N. Koudas, S. Muthukrishnan, and D. Srivastava. Approximate string joins in a database (almost) for free. In *Proc. 27th Int'l. Conf. on Very Large Databases*, pages 491–500, Roma, Italy, 2001.

[14] P. Gulutzan and T. Pelzer. *SQL-99 Complete, Really.* R&D Books, 1999.

[15] D. Guseld. *Algorithms on Strings, Trees and Sequences: Computer Science and Computational Biology.* Cambridge University Press, 1997.

[16] M. Hernandez. A generalization of band joins and the merge/purge problem. Technical Report CUCS-005-1995, Columbia University, February 1996.

[17] M. Hernandez and S. Stolfo. The merge/purge problem for large databases. In *Proceedings of the ACM SIGMOD International Conference on Managemnet of Data*, pages 127–138, May 1995.

[18] M. Hernandez and S. Stolfo. Real-world data is dirty: Data cleansing and the merge/purge problem. *Data Mining and Knowledge Discovery*, Vol. 2, No. 1:9–37, 1998.

[19] D. S. Hirschberg. Algorithms for the longest common subsequence problem. *Journal of the ACM*, 24:664–675, 1977.

[20] M. L. Jarke, M. Vassiliou, and P. Vassiliadis. *Fundamentals of data warehouses.* Springer, 2000.

[21] R. Kimball. Dealing with dirty data. *DBMS online*, September 1996. Available from `http://www.dbmsmag.com/9609d14.html`.

[22] K. Kukich. Techniques for automatically correcting words in text. *ACM Computing Surveys*, pages 24(4):377–439, 1992.

[23] L. V. S. Lakshmanan, F. Sadri, and I. N. Subramanian. SchemaSQL - a language for interoperability in relational multi-database systems. In *Proc. 22nd Int'l. Conf. on Very Large Databases*, pages 239–250, Mumbai, 1996.

[24] M. L. Lee, T. W. Ling, and W. L. Low. Intelliclean: A knowledge-based intelligent data cleaner. In *Proceedings of the sixth ACM SIGKDD international conference on Knowledge discovery and data mining*, pages 290–294, 2000.

[25] M. L. Lee, T. W. Ling, and W. L. Low. A knowledge-based framework for intelligent data cleansing. *Information System Journal - Special Issue on Data Extraction and Cleaning*, 2001.

[26] M. L. Lee, H. J. Lu, T. W. Ling, and Y. T. Ko. Cleansing data for mining and warehousing. In *Proceedings of the 10th International Conference on Database and Expert Systems Applications (DEXA)*, pages 751–760, 1999.

[27] Infoshare Limited. Best value guide to data standardizing. *InfoDB*, July 1998. Available from `http://www.infoshare.ltd.uk`.

[28] A. Marzal and E. Vidal. Computation of normalized edit distances and applications. In *IEEE Transactions on Pattern Analysis and Machine Intelligence*, pages 15(9):926–932, 1993.

[29] DataCleanser DataBlade Module. `http://www.informix.com/informix/products/options/udo/datablade/dbmodule/edd1.htm`.

[30] A. E. Monge. Matching algorithm within a duplicate detection system. In *IEEE Data Engineering Bulletin*, volume 23(4), December 2000.

[31] A. E. Monge and C. P. Elkan. An efficient domain-independent algorithm for detecting approximately duplicate database records. In *Proceeding of the ACM-SIGMOD Workshop on Research Issues on Knowledge Discovery and Data Mining*, Tucson, AZ, 1997.

[32] L. Moss. Data cleansing: A dichotomy of data warehousing? *DM Review*, February 1998. Available from `http://www.dmreview.com/editorial/dmreview/print_action.cfm?EdID=828`.

[33] Dictionary of Algorithms and Data Structures. http://www.nist.gov/dads/.

[34] X. Y. Qi, S. Y. Sung, C. Lu, Z. Li, and P. Sun. Field similarity algorithm. In *Sixth International Conference on Computer Science and Informatics*, pages 432–436, Durham, NC, USA, March 2002.

[35] E. Rahm and H. H. Do. Data cleaning: Problems and current approaches. In *IEEE Data Engineering Bulletin*, volume 23(4), December 2000.

[36] V. Raman and J. M. Hellerstein. Potter's wheel: An interactive data cleaning system. In *Proc. 27th Int'l. Conf. on Very Large Databases*, pages 381–390, Rome, 2001.

[37] G. Riley. A tool for building expert systems, 2002. Available from `http://www.ghg.net/clips/CLIPS.html`.

[38] A. Silberschatz, M. StoneBraker, and J. Ullman. Database research: Achievements and opportunities into the 21st century. In *SIGMOD Record (ACM Special Interest Group on Management of Data)*, page 25(1):52, 1996.

[39] T. F. Smith and M. S. Waterman. Identification of common molecular subsequences. *Journal of Molecular Biology*, pages 147:195–197, 1981.

[40] R. E. Tarjan. Efficiency of a good but not linear set union algorithm. *Jouanal of the ACM*, 22(2):215–225, 1975.

[41] E. Ukkonen. On-line construction of suffix trees. *Algorithmica*, 14(3):249–260, 1995.

[42] R. Wagner and M. Fisher. The string to string correction problem. *Jouanal of the ACM*, 21(1):168 173, 1974.

[43] R. Y. Wang, M. P. Reddy, and H. B. Kon. Towards quality data: An attribute-based approach. *Decision Support Systems*, 13, 1995.

CLUSTERING AND INFORMATION RETRIEVAL (pp. 261-298)
W. Wu, H. Xiong and S. Shekhar(Eds.)

A Science Data System Architecture for Information Retrieval

Daniel J. Crichton
Jet Propulsion Laboratory
4800 Oak Grove Drive
Pasadena, California 91109
E-mail: Daniel.J.Crichton@jpl.nasa.gov

J. Steven Hughes
Jet Propulsion Laboratory
4800 Oak Grove Drive
Pasadena, California 91109
E-mail: Steve.Hughes@jpl.nasa.gov

Sean Kelly
Independent Consultant
1672 Hope Drive Unit 1817
Santa Clara, California 95054-1718
E-mail: Sean.Kelly@jpl.nasa.gov

Contents

1 Introduction

Science research generates an enormous amount of data that is located in geographically distributed data repositories. The data generated by these efforts are often captured and managed without reference to any standard principles of information architecture. Interoperability and efficient search and retrieval of data products across disparate data systems is difficult because users are often required to connect to each individual data system and deal with dissimilar and often unfamiliar interfaces and semantics. It makes the development of software systems that work across organizational and disciplinary boundaries challenging if the organizing principles that construct the information architecture are not explicitly defined. Clustering data results across multiple information systems is challenging without a system architecture that provides both the data and distributed systems architecture and standards.

At the National Aeronautic and Space Administration's (NASA) Jet Propulsion Laboratory (JPL), we've been researching distributed data architectures to support data collaboration with space science and biomedical research through an initiative called the Object Oriented Data Technology (OODT) project. Our project researched and developed a metadata driven framework allowing for the discovery and exchange of data resources and products across widely distributed data repositories referred to as data nodes. OODT identified several key project goals for the framework [4] which include:

1. encapsulating data nodes to hide uniqueness
2. use of metadata for all messages exchanged between distributed services
3. a standard data dictionary for describing data resources
4. use of ubiquitous interfaces across multiple data systems that provide interoperability via a common query mechanism
5. the establishment of a standard data model for describing any data resource irregardless of its location

Clustering of multi-discipline data systems is challenging due to incompatible and often undefined domain data models. Domain data models provide the semantic data architecture necessary for correlating results from multiple data systems. The OODT project has worked to establish data "communities" which allow domain models to be clustered around the data architectures created for these communities. For example, within NASA's Office of Space Science, there exist hundreds of data systems containing several data sets and products across planetary, astrophysics and space physics missions and experiments. Each of these form communities that define their own data architecture focused on providing the semantic standards necessary for defining and correlating data results across research efforts.

While the data architecture is essential in providing a standard language for describing data resources, the distributed systems architecture is needed in order to provide the technology infrastructure to allow data to be shared between distributed data nodes. Several distributed systems standards and frameworks exist that allow data across a network to be exchanged between two points. The OODT framework uses a distributed object standard named the Common Object Request Broker Architecture (CORBA) to exchange metadata and data products using a common data format described with the Extensible Markup Language (XML) [1]. The distributed systems infrastructure allows for the publication of metadata resources that describe the data products and their associated locations that exist within a data community.

Two key efforts leveraging the OODT framework include NASA's Planetary Data System (PDS) and the National Cancer Institute's Early Detection Research Network (EDRN). Both of these efforts define a domain-dependent data architecture including a common data model and data dictionary to describe data products that are specific to those efforts, however, many of the participating data nodes that form the collaboration are implemented to local standards in terms of their databases and computing platforms. In addition, the data that is collected and distributed across these efforts is located at participating research institutions based on the institution's expertise to generate and analyze the data. This chapter will focus on the development of a common data and distributed systems approach to building a solution that allows for the clustering and retrieval of information to support science research within these domains.

1.1 Related Work

Similar work takes the form of the ISAIA [2] project, originally proposed in 1999 to develop an interdisciplinary data location and integration service for space science.

Building upon existing data services and communications protocols, it was envisioned that this service would allow users to transparently query hundreds or thousands of WWW-based resources from a single interface. The service would collect responses from various resources and integrate them in a seamless fashion for display and manipulation by the user.

The ISAIA pilot study was influential in shaping the science goals, system design, metadata standards, and technology choices for the virtual observatory. The ISAIA pilot project also helped to cement working relationships among the NASA data centers, US ground-based observatories, and international data centers and was formed as a collaborative effort between thirteen institutions that provided data to astronomers, space physicists, and planetary scientists.

Much effort went into defining metadata standards and exploring the means by which such standards should be expressed. We developed the concept of profiles, i.e., groups of terms, their associated values or value ranges, and the relationships between terms that could be used to describe data collections, information services, and individual observations or images. We also characterized profiles as resource profiles (used to describe a service), query profiles (used to describe how a query to a service should be expressed) and response profiles (used to describe the information returned from a service). These profiles share many terms and relationships, though a response profile, for example, might return considerably more information about a data set than the associated query specifies.

2 Data Architecture

A simple definition for information in Webster's New Collegiate Dictionary is "a ... character representing data [10]." For example, the character 7 represents data. However, besides implying that something has been counted, there is very little knowledge communicated by this single character. By providing descriptors of the data, also known as data about data or alternately metadata, more knowledge can be communicated. For example, if the character 7 is described as a value of latitude, the knowledge communicated is more complete and useful.

In most domains whether medical research, space science, or engineering precision in both the description of a value as well as a description of the description is critical for the communication of knowledge. For example, what is latitude? This implies that descriptors of the metadata, also known as data about metadata or alternatively meta-metadata, is needed. For example, latitude could be described using a definition, the data type of its values, and the units in which the values were measured. This would allow the value 7 to be more precisely described as the integer value of an angle measured in units of degrees that a straight line, passing through both a point at the center and a point on the surface of a sphere, subtends with respect to the equatorial plane of a sphere. This level of precision communicates even more knowledge.

Even more knowledge can be communicated by describing the relationships between data. For example, by relating our value of latitude to observed bodies and types of maps, it could be placed in context as a latitude on Mars in a sinusoidal map projection.

The above discussion illustrates that the collection of data and relationships at several levels is key to the communication of knowledge. It also proves the truth of the assertion "that the metadata remain attached to the data or the data becomes meaningless and unusable [11].". For example, a spacecraft camera image is just a pretty picture unless calibration data is available for reducing the data captured by the camera to physical parameters for use by scientists.

The key components of the OODT data architecture are the standards used for collecting, organizing, and managing data and their relationships. In the following section a hierarchy of standards will be described, starting at the base of the hierarchy with meta-metadata or the set of descriptors needed to describe descriptors.

2.1 Descriptors for Describing Metadata

The most basic component of a data architecture is the set of descriptors needed to describe descriptors. These descriptors are also called meta-attributes or meta-metadata and are used to systematically create a dictionary of terms used in an area of interest, also called a domain.

For example, the Webster's New Collegiate Dictionary provides an explanatory chart that illustrates descriptors used in that dictionary, such as main entry, definition, and synonym cross-reference. These descriptors are used to concisely and consistently describe entries in the dictionary so that the entries can be used intelligently in an English sentence. For example, the main entry "information" is described using the definition "a ... character representing data" and is cross-referenced to **facts** and **data** as synonyms. In a similar manner, the descriptors of entries in a medical research or space science dictionary must be concisely defined to intelligently use the dictionary entries in their respective domains.

ISO/IEC 11179 [8] is a framework for the specification and standardization of data elements. It provides a base set of descriptors needed to describe descriptors. As an international standard, it provides a common basis for data element definition and classification across many areas of interest. The specification defines four data element categories namely identifying, definitional, representational, and administrative as presented in Table 1. Note that the terms descriptor and data element are synonymous for this discussion and that the term attribute in the specification is a descriptor of a descriptor.

Attribute Category	Name of data element attribute	Obligation
Identifying	name	M
	identifier	C
	version	C
	registration authority	C
	synomymous name	O
	context	C
Definitional	definition	M
Representational	datatype of data element values	M
	maximium size of data element values	M
	minimum size of data element values	M
	permissible data element values	M
Administrative	comments	O

Table 1: ISO/IEC 11179 Basic Attributes

The identifying category is used for the identification of a data element. For example, the attribute identifier uniquely identifies a data element within an area of interest. The definitional category is used to describe the semantic aspects of a data element and consists of a textual description that communicates knowledge about the data element that typically is not captured by any of the basic attributes. The relational category describes associations among data elements and/or associations between data elements and classification schemes, data element concepts, objects, or entities. For example relating the latitude to observed body provides critical information about how latitude is to be interpreted. The representational category describes representational aspects of data element such the list of permissible data values and their type. Finally the administrative category provides management and control information.

The obligation column in the table designates whether an attribute is mandatory (M) or always required, conditional (C) or required under certain conditions, and optional (O) or simply allowed.

The application of a general specification such as ISO/IEC 11179 to a specific domain requires specialization. For example the attribute "datatype of data element values" is defined as "A set of distinct values for representing the data element value." Best practices would suggestion that it be constrainted by adopting a specific standard that provides an enumeration of datatypes.

2.2 Global Descriptors for Describing Electronic Resources

With the advent of the web and the resulting explosion of electronic resources available for online access, there was a compelling need for standard descriptors for the multitude of electronic resources on the web. Of course, since information contained in these electronic resources span the breadth of human knowledge, the set of standard descriptors that span all electronic resources must be very general and limited in number.

The Dublin Core [5] initiative specifically addresses this issue and recommends a list of 15 data elements or descriptors as presented in Table 2.

Validating the previous discussion, the ISO/IEC 11179 attributes are used by the Dublin Core Initiative to define the Dublin Core descriptors. As stated on the Dublin Core website each Dublin Core element was defined using ten ISO/IEC 11179 [ISO11179] descriptors.

These include:

- Name: The label assigned to the data element

- Identifier: The unique identifier assigned to the data element

- Version: The version of the data element
- Registration Authority: The entity authorised to register the data element
- Language: The language in which the data element is specified
- Definition: A statement that clearly represents the concept and essential nature of the data element
- Obligation: Indicates if the data element is required to always or sometimes be present (contain a value)
- Datatype: Indicates the type of data that can be represented in the value of the data element
- Maximum Occurrence: Indicates any limit to the repeatability of the data element
- Comment: A remark concerning the application of the data element

Six of the ten descriptors however were common to all of the Dublin Core elements:

Version	1.1
Registration Authority	Dublin Core Metadata Initiative
Language	en
Obligation	Optional
Datatype	Character String
Maximum Occurrence	Unlimited

2.3 A Schema for Describing Domain Resources

The Dublin Core descriptors are limited in number and by definition very general. When used to search a large number of electronic resources across many repositories, they will will typically produce either a very small or very large result set. Developed to describe resources across all possible domains, they are limited in their ability to partition the search space into manageable subsets. For example, a search using Identifier would typically resolve to one resource while a search using Type would typically resolve to a large number of resources.

Even when specialized for use in a particular domain the Dublin Core descriptors will still remain limited in their ability to resolve queries. For example the descriptor Date might be specialized to require that its value conform to the ISO 860 standard. This will help to make searches involing date more exact however unless the user knows the approximate date associated with the resource, date can not be used. The solution to this problem is the use of domain specific descriptors.

The Dublin Core descriptor Subject allows the use of domain terminology to describe resources. "Typically, a Subject will be expressed as keywords, key phrases or classification codes that describe a topic of the resource. Recommended best practice is to select a value from a controlled vocabulary or formal classification scheme." [5] For the data architecture, a controlled vocabulary is used but captured separately from the Dublin Core descriptors. The descriptor Subject can either be ignored or used for keywords.

In the OODT data architecture, an electronic resource is described by a profile. A schema for the profile is provided in Figure 1, and has three groups of descriptors: profile descriptors, resource descriptors, and descriptors from the domain controlled vocabulary, called the profile elements. The first section, the profile descriptors, simply describes the profile itself and contains system level attributes such as profile identifier, type, and status.

The second section, the resource descriptors, generically describes the resource using the Dublin Core descriptor set. All descriptors are allowed but only Identifier is required. Additional resource descriptors have been added to identify the resource's local domain, classification, and the location.

Finally, the profile element section provides domain specific descriptors for the resource. These descriptors are typically extracted from a domain data dictionary.

For example, a profile to describe a web-based planetary image catalog, would include the Identification, Title, and location of the catalog. The profile element section would include planetary science domain descriptor such as as planet, spacecraft, and instrument names.

The profile, as shown in Figure 1, was implemented using the Extensible Markup Language [1]. XML provides a number of advantages including simplicity, expressiveness to HTML, industry adoption, and flexibility. The XML Profile is an important part of the data architecture for describing science data resources and will be described in further detail in this chapter.

2.4 Data Models to Describe Information Resources

The key organizing principle within a data architecture is the data model. A data model is used to formally describe the data in terms of its structure and relationships. Examples of data models include the Planetary Science data model [12] for the Planetary Data System (PDS) data archive, the

```
<!ELEMENT profiles
  (profile*)>

<!ELEMENT profile
  (profAttributes,
   resAttributes,
   profElement*)>

    <!ELEMENT profAttributes
      (profId, profVersion?, profType,
       profStatusId, profSecurityType?, profParentId?, profChildId*,
       profRegAuthority?, profRevisionNote*, profDataDictId?)>

    <!ELEMENT resAttributes
      (Identifier, Title?, Format*, Description?, Creator*, Subject*,
       Publisher*, Contributor*, Date*, Type*, Source*,
       Language*, Relation*, Coverage*, Rights*,
       resContext+, resAggregation?, resClass, resLocation*)>

    <!ELEMENT profElement
      (elemId?, elemName, elemDesc?, elemType?, elemUnit?,
       elemEnumFlag, (elemValue* | (elemMinValue, elemMaxValue)),
       elemSynonym*,
       elemObligation?, elemMaxOccurrence?, elemComment?)>
```

Figure 1: XML DTD for a Profile

Early Detection Research Network Common Data Element iniative, and the Dublin Core elements [5] for describing electronic resources on the Web.

2.4.1 Metamodels

A metamodel is used to describe a data model. Often ambiguously also called "model," a metamodel prescribes or provides the syntax and semantics for creating and documenting data models as illustrated in Figure 2. An example of a metamodel is the Entity-Relationship (E-R) model. Metamodels provide a set of normative or prescribing standards for capturing the meaning, relationships and behavior of data within the data space. They often use graphical notation to define the objects and relationships that exist within the data space.

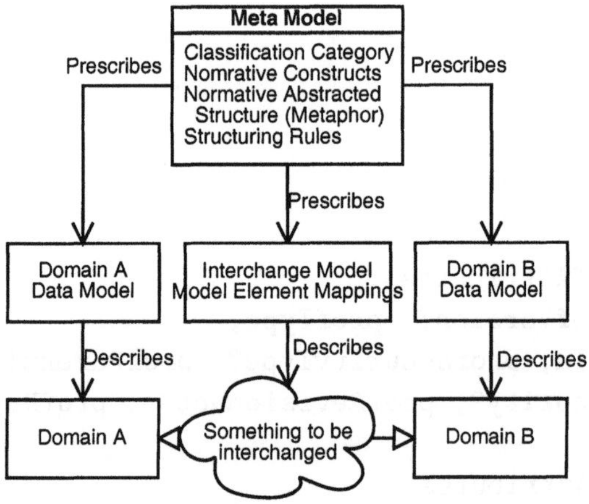

Figure 2: Relationship of Domain Data and Meta Models

In the example illustrated in Figure 3, the PDS used the E-R metamodel to define how images of a planet are related to the spacecraft instrument that captured them. In the PDS data model, an entity called "data set" was first defined to describe a collection of images and then an "instrument" entity was defined to provide both summary and detailed descriptions about an instrument's functionality. A one-to-many relationship, "produces," was then defined between the instrument and data set entities to allow one instrument to be related to the many data sets that it could produce.

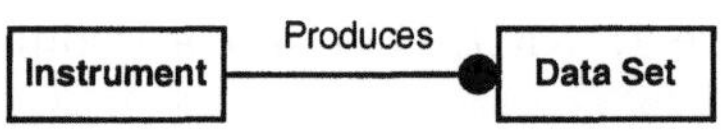

Figure 3: PDS Instrument-Spacecraft Relationship

The Entity-Relationship (E-R) model is only one of many metamodels. Metamodels can be classified as either object-based or record-based. The object-based metamodels include E-R and object-oriented. The record-based include relational, networked, and hierarchical. Each of these metamodels has its own syntax, semantic constructs, and best practices for their application within information systems. Within an enterprise, each of these has probably been used at least once and has probably been applied with varying degrees of compliance to the specifications. In fact, only the E-R and relational metamodels come close to having what can be considered well-defined and widely accepted specifications.

2.4.2 Search and Retrieval Data Architectures

To provide efficient and effective search and retrieval, the technology architecture needs to conceptually represent the data architecture to the user. To achieve this, information system developers typically rely on user interfaces to conceptually represent the data model in a user friendly and intuitive manner. For example, because planetary science data sets are related to target bodies, an intuitive user interface might start with a request for a target body, possibly represented by an icon depicting a planet or satellite. The user could then be guided to select a spacecraft that flew by the planet to further constrain the search. The development of the technology architecture and specifically user interfaces is relatively easy if there is a well-defined data architecture.

Search and retrieval across more than one information system is not much more difficult if the information systems belong to the same domain and use the same data architecture. However, this is seldom the case. Often the need arises to search across multiple domains that have little in common except that the user believes that some common relationship exists. For example, the Hubble space telescope is considered part of the astrophysics space science domain, focusing on objects outside our solar system. However, the telescope is often used to capture images of planets and their satellites. So it is conceivable for a planetary scientist to request a search across both the astrophysics and planetary domains for images of Jupiter. However, the data architecture of the two domains are entirely different. The astrophysics domain works primarily with extra-solar objects and assumes stationary target bodies that have the celestial sphere as a frame of reference. It also allows target bodies to have many names. The planetary domain, however, deals with orbiting target bodies and fly-by spacecraft and so must deal with many related frames of reference. In addition, each target body is allowed only one standard name. So, a simple query to search for all images of Jupiter across both domains would first require collecting all the identifiers used for the planet Jupiter. Finding images of Jupiter by location using a simple cross-domain query is currently impossible.

2.4.3 Metadata Registries

Any solution to the interoperability problem involves identifying and describing commonalities across domain data models. This typically requires that the individual data models be normalized at the metamodel level, for example expressing the data models using ISO/IEC 11179 specifications.

The metadata registry uses ISO/IEC 11179 specifications for capturing domain vocabularies and data model and ontological relationships. In particular the relational category is used for entity relationships. Data models are normalized once they have been imported and can then be analysed by domain experts to find and document commonalities. The following scenario illustrates this approach. After importing the astrophysics and planetary data models, the astrophysics and planetary terms for target bodies, TARGET_NAME and OBJECT_ID, would be identified as two terms for the same concept. They could then be described as synonyms since for example there is only a simple syntactical difference between TARGET_NAME=JUPITER and OBJECT_ID=JUP.

More significant semantic differences can be handled using ontological concepts. For example both the planet Jupiter and the star Vega are target bodies but are targeted for study in different frames of reference. The common ontological concept of "target bodies of interest" can first be defined and then the terms planet and star related to it. The resulting relationship spans two domains and enables some level of interoperability.

11179 Attributes	Dublin Core Element
Name	Title
Identifier	Title
Definition	A name given to the resource.
Name	Creator
Identifier	Creator
Definition	An entity primarily responsible for making the content of the resource.
Name	Subject and Keywords
Identifier	Subject
Definition	The topic of the content of the resource.
Name	Description
Identifier	Description
Definition	An account of the content of the resource.
Name	Publisher
Identifier	Publisher
Definition	An entity responsible for making the resource available
Name	Contributor
Identifier	Contributor
Definition	An entity responsible for making contributions to the content of the resource.
Name	Date
Identifier	Date
Definition	A date associated with an event in the life cycle of the resource.
Name	Resource Type
Identifier	Type
Definition	The nature or genre of the content of the resource
Name	Format
Identifier	Format
Definition	The physical or digital manifestation of the resource.
Name	Resource Identifier
Identifier	Identifier
Definition	An unambiguous reference to the resource within a given context.
Name	Source
Identifier	Source
Definition	A Reference to a resource from which the present resource is derived.
Name	Language
Identifier	Language
Definition	A language of the intellectual content of the resource.
Name	Relation
Identifier	Relation
Definition	A reference to a related resource.
Name	Coverage
Identifier	Coverage
Definition	The extent or scope of the content of the resource.
Name	Rights Management
Identifier	Rights
Definition	Information about rights held in and over the resource.

Table 2: Dublin Core Descriptors

3 Distributed Services

The OODT framework that implements the dynamic data model described in Section 2 consists of a set of cooperating, distributed peer software components. Although we've currently implemented the software with a client-server communications substrate, the design resembles a peer-to-peer (P2P) network, and we have plans to transition the communications substrate to a P2P implementation, probably based on JXTA [6].

The distributed services of the OODT framework consist of several major components that interoperate in such a way that they implement the metadata (profile) and data (product) model. They make it possible for researchers to locate data independent of its physical location and format, and to correlate data by leveraging metadata.

3.1 Distributed Framework Communication

OODT is a distributed system, wherein components may be dispersed geographically across a standard TCP/IP network, such as the Internet. Connectivity between components utilizes the Internet Inter-ORB Protocol (IIOP) for CORBA components.

Components themselves support plug-ins that perform the work of querying for metadata and data. In this way, the OODT software is a *framework* [7]. In a traditional library API approach, application programmers use library classes and methods to develop a complete program. In the framework approach, application programmers extend API classes and implement API interfaces that plug into and become part of the framework. Generally, applications based on frameworks are easier to write, though more specialized (and therefore limited) to a set task. More work necessarily falls upon the developers of the framework to support all sorts of plug-ins.

OODT's framework provides three major components:

- **Profile servers** serve scientific metadata and can tell whether a particular resource can provide an answer to a query.
- **Product servers** serve data products in a system-independent format.
- **Query servers** accept profile and product queries and traverse the network of profile and product servers, collecting results.

The servers communicate with each other by passing XML [1] messages to each other over CORBA method calls.

The query server is the starting point for all end-user activity with the system. Using CORBA, Java, or HTTP, researchers can make profile queries to determine what resources exist to satisfy a query. For resources that are product servers, researchers can make product queries to retrieve data.

3.2 Common Design Elements

The OODT framework servers exhibit some common design elements. They all advertise a remotely accessible program interface using the CORBA Interface Definition Language. Clients may contact profile and product servers directly in this way instead of using the query server's interface. Typically, though, clients use the query server as a starting point into the system. The query server emulates a gateway into a P2P space using CORBA client/server method invocations to profile and product servers. Figure 4 demonstrates a typical deployment of the OODT software. In the diagram, one system runs a query server and a root profile server. The root profile server contains profiles that describe two other profile servers. One of these other profile servers contains profiles that describe resources within a product server that serves images; this product server retrieves resources from a database. The other profile server contains profiles that describe resources in a document collection with a web front end.

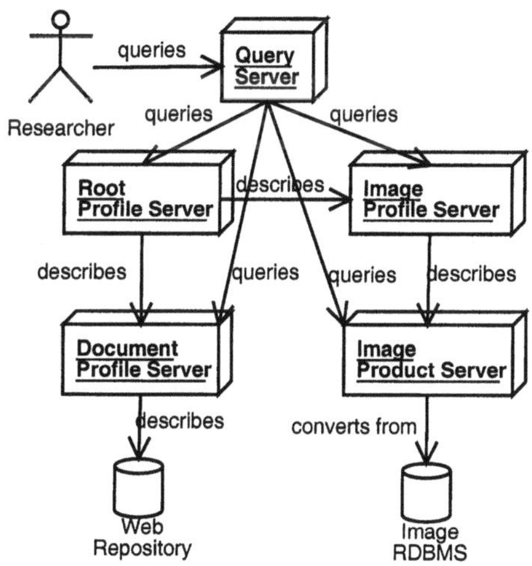

Figure 4: Typical deployment of OODT

Each profile and product server use a set of customizable backend implementations in order to process queries. For example, a profile server may retrieve profile metadata from a flat comma-separated values file, from an XML database, from a relational database, from a document catalog, and so on. Similarly, a product server could retrieve images stored in a proprietary format as BLOBs in a relational database and return them in standard PNG format. Both profile and product servers specify the interfaces for interchangeable backends; by creating implementations that conform to those interfaces, we can provide different and user-customizable behavior that's compatible with the OODT framework. These interfaces are specified using Java's interface mechanism. Therefore, backends are classes that implement the interface. At run time, profile and product servers consult the system properties to determine which backend classes to load, instantiate, and install as backends.

In order to handle queries, both profile and product servers manipulate an identical query structure we call the XML Query. (It's called such only because it can be expressed in XML.) Representable as both an XML document and as an object of a Java class, the XML Query encapsulates the user's query in a query-language–independent fashion along with any results retrieved so far.

3.3 Profile Servers

As described in Section 2, profiles are metadata descriptions of resources; that is, they "profile" a resource by describing its inception and composition. Profile servers serve profiles. They manage a collection of profiles, providing a way to query and update that set. Profile servers answer the question, "Where can I go to find out about X?"

All profile operations that arrive at a profile server are delegated to the server's current backend implementation. Figure 5 demonstrates the delegation.

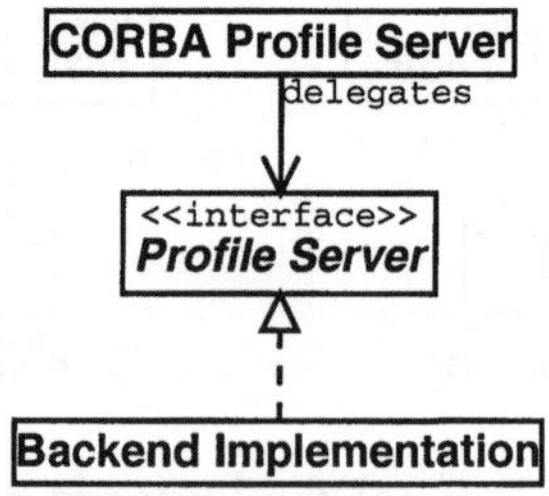

Figure 5: Delegation architecture of a profile server

The Profile Server is a Java interface; therefore, backend implementations are Java classes that implement the interface. In general, two kinds of backends can be developed:

- **Static** profile servers serve profiles that exist statically. These are profiles that typically describe long-lived resources that do not change; thus, the profiles themselves rarely need change.

- **Dynamic** profile servers create profiles on-the-fly in response to profile queries. Such profiles may describe ephemeral resources, or long-lived resources for which having static profiles would be onerous. For example, having profiles for millions of different datasets that vary in only small ways would require too much disk space or memory to maintain; a dynmaic profile server can synthesize profiles for such resources based on one copy of the nonchanging information.

3.3.1 Profile Queries

A profile server's primary responsibility is to provide a way to run a query against the server's set of profiles. As mentioned before, users may access a profile server directly via its CORBA interface. More commonly, though, queries enter the system through the query server, which directs them transparently to and along graphs of appropriate profile servers.

Upon receiving a query, the profile server's backend interprets the XML Query passed in a way appropriate to the implementation. For example, a backend that stores information in a relational database may convert the XML Query into an SQL query. For each matching profile, the backend constructs a Document Object Model (DOM) document describing the profile, in XML, and adds them to a single DOM document containing all matching profiles.

Consider an example profile server that manages a set of profiles that describe image maps of a body. Each image map is a two dimensional image that places at each latitude and longitude pixel a physical measurement. A profile for a thermograph map would have a profile that describes the range of latitude and longitude covered by the map and the the range of temperatures. A query for resources that contained temperature, or a query for a specific temperature range, or a query that combined temperature and position could all match this profile. The profile server serving that profile would answer these queries by examining the request and returning the matching profile in DOM format.

3.3.2 Profile Management

In addition to searching a set of profiles, the profile server interface describes methods to manage the profiles. Profile management includes adding new profiles, adding new versions of profiles, removing profiles, iterating over the set of profiles, retrieving a single profile, and clearing the set of profiles. By implementing the management methods, the backend provides a way for remote administration of the metadata managed by a profile server.

Although the Java language requires that classes that implement an interface implement **all** the methods listed or else be abstract, not all management methods are appropriate for certain kinds of profile servers. For example, a profile server whose backend synthesizes ephemeral profiles upon receiving queries has no set of profiles to manage. In this case, we deem it sufficient for each management method to throw an unsupported operation exception.

3.3.3 Current Profile Backends

Thanks to the plugin architecture, we've been able to experiment with several backend implementations for static profile retrieval.

- **Lightweight Profile Server.** The Lightweight Profile Server is called lightweight because it doesn't rely on any external products or components. It reads in a set of profiles from a static XML file, constructs DOM trees for each one in memory, and handles queries by traversing the trees. It's fairly fast, but memory intensive. It's not ideal for handling large numbers of profiles, as a result.

- **Oracle Profile Server.** The Oracle Profile Server backend uses an Oracle 8 relational database to store profile information. Instead of storing complete XML documents, the Oracle backend decomposes each XML-based profile into a number of rows into a number of tables. Querying involves constructing an appropriate SQL query and creating matching profiles in XML on-the-fly.

In addition, we've implemented a number of dynamic profile servers for highly specific applications. These include:

- Generating profiles for documents stored in Xerox's Docushare document management system. Incoming profile queries generate queries into Docushare's WebDAV-like API; the server uses the results to synthesize profiles that describe matching documents.

- Generating profiles for documents stored in a proprietary document collection. Incoming queries search the document indexes; results are used to synthesize profiles describing matching documents.

3.4 Product Servers

Product servers exist to provide a way to retrieve specific data products. Like their profile server kin, product servers accept the same XML Query struture. Instead of adding matching profiles to the results section of the query, product servers add matching products. Data products in this sense can be individual data granules, datasets, or collections of datasets, depending on the backend implementation in the product server and the way it handles queries and results.

As with profile servers, product servers delegate requests into a specific backend to perform the actual query. Unlike profile servers, there can be multiple backends in a single product server. Each backend concurrently gets a chance to handle the query and add any matching results. Figure 6 shows the class architecture.

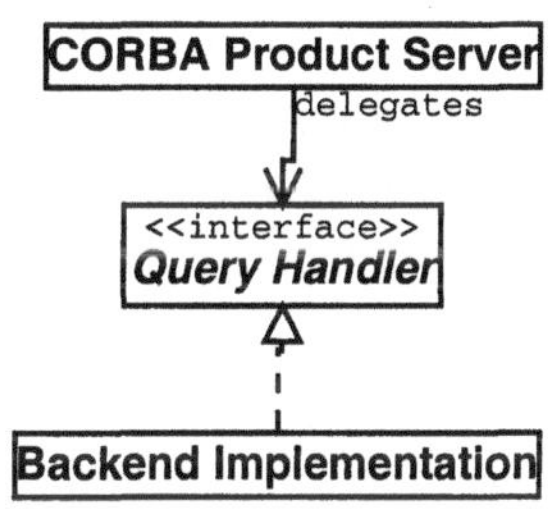

Figure 6: Delegation architecture of a product server

The backend interface is called a Query Handler and is deployed as a Java interface. Therefore, backend implementations are concrete classes that implement the interface. The Query Handler interface is far simpler than the Profile Server interface, though: there is a single method that must be implemented. This method accepts an XML Query object describing the user's request. Any matching products get inserted into the XML Query object, which is then returned.

3.4.1 Product Formats

The primary responsibility of a product server is to return products in standard formats. To do so, the product server's backends must be capable of handling Internet standard MIME [9] types, and converting data products from their native storage formats into standard MIME formats.

When constructing a query, the user may indicate preferred MIME types. For example, a user wanting PNG images may list `image/png` as the only acceptable MIME type. A user preferring PNG images but willing to have JPEG images would list `image/png`, `image/jpeg` in that order. A user preferring PNG images but willing to accept any image type would list `image/png`, `image/*`. If the user doesn't specify a MIME type when creating the XML Query object, the software generates a default list of acceptable MIME types, namely `*/*`, meaning that any type is acceptable.

The XML Query's result section indicates the MIME type of each result, and product server's query handlers are obligated to indicate the MIME type of each result. For example, a product server that generates image maps from contour data may return products in the `image/jpeg` format. A product server that retrieves tabular data from a database may return products in the `text/tab-separated-values` format.

The OODT framework software includes an automatic mechanism for encoding products between software components within the XML document that's driven by MIME type. For those MIME types like `image/png` or `application/ms-word` that are binary data get encoded into base-64 [9] text. Textual types like `text/tab-separated-values` and `text/html` get no special encoding and are just placed within an XML CDATA section.

3.4.2 Invoking Query Handlers

The framework software of the OODT product server code initializes each query handler when the product server is started. The product server software consults a system property for a list of class names; it loads each class, creates one instance, and attempts to install the instance as a query handler. In this way, you can change the query handlers installed at a product server at startup time by changing the system property.

When queries arrive at the product server, the product server invokes each instantiated query hander's query method. Each handler may add as many products as it wishes, but it should respect the user's maximum product cap (specified during construction of the XML Query object).

Every product added to the XML Query object receives a string identifier to differentiate it from other matching products in the result. Currently, there is no enforcement of uniqueness to this string. Typical usage might be to assign a long term, persisitent identifier such as an OID for products, and use UUIDs for other products, but no query handler has implemented such features at this time. To resolve collisions, the user may iterate over every product in the result set in the XML Query object.

3.5 Query Servers

Query servers service queries. They're the point of entry into an OODT framework installation. Query servers contain the algorithms necessary to traverse the logical P2P model in its physical client/server deployment, executing queries at appropriate servers and gathering results. Query servers also simplify the interaction with the user, who is freed from the knowledge of accessing the CORBA interfaces of profile servers and product servers. Users instead call upon a query server for all profile and product interaction using either HTTP, a Java API, or CORBA.

The HTTP interface is ideal for creating links to products within web pages. It's also useful for languages that don't have a CORBA interface: most languages have ready APIs for accessing information via HTTP, and you can leverage the HTTP interface within analysis or other programs to retrieve metadata and data. We implemented the HTTP interface by calling upon the Java API; the interface resides within a servlet in any servlet-capable web server.

Java users will want to use the Java API. Other users who do not wish to use the HTTP interface but do have access to CORBA (such as C/C++ users) can call the CORBA interfaces.

3.5.1 HTTP Interfaces

We provide two different HTTP interfaces:

- **Generic query interface.** This interface requires that the caller be able to construct and parse XML documents. It provides full access to a Query Server for the retrieval of metadata (profiles) and data (products).

  To use this interface, the caller constructs an XML Query and sends it to the server in serialized (string) XML form, along with the search type (profile or product) and the name of the server to receive the query. The return value is an XML document (MIME type `text/xml`) with any results.

- **Product query interface.** This simpler interface enables access to products only—profiles cannot be retrieved. To use this interface, the caller sends in a keyword query string and an optional list of desired

MIME types. The servlet constructs a corresponding XML Query object and executes the query. It returns the first matching product in its MIME type as a result. In the case there are multiple matching products, the user may pass in a result ID string and just that one will be returned.

Using this interface it is trivial to create web pages whose graphics (for example) are retrieved through the product service.

3.5.2 Typical Usage

End users usually will not have a priori knowledge of what profile and product servers exist and by what CORBA network names they're known and accessed. And there usually isn't any need to bother asking system administrators for such information. The OODT framework is typically bootstrapped in such a way that there's a root profile server that contains metadata descriptions of other profile servers. These other profile servers may in turn describe yet more profile servers and/or product servers and other resources.

Using this network architecture, a user just sends in a profile query to the default profile server. The query server will crawl the network to gather results and return a list of known product servers that can provide the sought product. The user can then submit a second query (a product query) targeted at a specific product server to retrieve the sought data. Figure 7 demonstrates how this usually works.

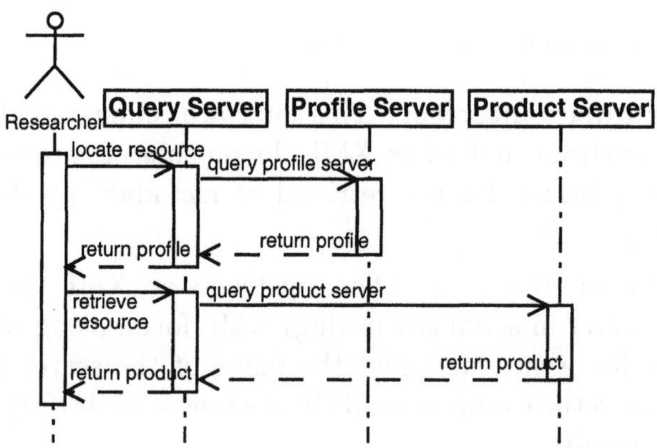

Figure 7: Typical interaction with the OODT framework

4 Applications

This section describes two specific applications of the OODT data architecture and distributed services.

4.1 Planetary Data System

NASA's Planetary Data System (PDS) is an active science data archive managed by scientists for NASA's planetary science community. In operation since 1990, the archive currently has about five terabytes of data covering 30 years of solar system exploration. This archive will double in size within a year and will increase by a factor of 60 by 2005. The system is distributed, consisting of several science disciplines nodes, support nodes, and a central node.

The primary focus of the PDS has been the creation of a high quality long-term science data archive. Historically the PDS archives and distributes the data to the community on CD/DVD media. It also provides on-line search and retrieval capabilities through discipline-oriented interfaces at the nodes. For example, the PDS imaging node is responsible for imaging data products and provides on-line access to most of its imaging data products. The guiding philosophy was that domain scientists either receive a copy of the data on physical media or used the appropriate discipline node interfaces to access the data.

With the dramatic increase in data volumes for future missions, the creation of multiple copies of each physical volume for distribution is no longer viable. This suggests on-line search and retrieval as the primary means of data distribution. In addition, loosely coupled and domain oriented system development has resulted in heterogeneous systems that make interoperability difficult. Interoperability is now desired to support spatial/temporal access, science analysis, and knowledge discovery across the discipline nodes. Finally the number of point-to-point interconnections have increased exponentially as developers attempt to support interoperability by simply linking in new data nodes.

The key to addressing these issues is an information system architecture that distinguishes the data architecture from the technology architecture. The PDS data architecture, developed very early, has proven to be the single most important element in maintaining consistency across discipline nodes. The data architecture includes a data model, a scheme for collecting and associating metadata with science data products, and a peer review process for ensuring data and metadata validity. The wealth of metadata already existing in the archive can be leveraged to support interoperability since it is sufficient to identify and describe resources and provides the necessary search attributes for finding resources.

The Object Oriented Data Technology (OODT) framework provides the distributed systems infrastructure to interconnect the Planetary Data System's geographically distributed nodes. It operates concomitantly with existing PDS subsystems, supporting them as indepedent data nodes interoperating through a standards-based communications infrastructure. Product servers are being implemented at each node data repository to provide a common interface for retrieving the data products that are located at those nodes. The product servers allow the user to request data products in a variety if formats including raw, derived, and packaged. For example, the Mars Odyssey Neutron Spectrometer data product can be retrieved either as a ZIPed binary data product or an ASCII file.

Profile servers are being implemented for PDS data set and data product catalogs to provide a common interface for searching. For example, the PDS Central Node catalog contains an inventory of all data sets in the archive. A PDS data set profile server has been implemented for this catalog that returns XML profiles describing the data sets as well as providing the location of applications that either provide data set browse or retrieval capabilities. An example profile for a Mars Odyssey product is defined below in figure 8.

```
<?xml version="1.0"?>
<!DOCTYPE profiles SYSTEM "http://somehost.jpl.nasa.gov/dtd/prof.dtd">
<profile>
   <profAttributes>
     <profId>1.3.6.1.4.1.1306.2.104.10018791</profId>
     <profVersion>null</profVersion>
     <profType>profile</profType>
     </profAttributes>
   <resAttributes>
     <Identifier>ODY-M-HEND-EDR-2-V1.0+H0133</Identifier>
     <Title>Data_Set_Name: ODYSSEY-MARS-HEND-EDR-2-V1.0 + Product_Id:H0133</Title>
     <Description>Mars Odyssey Dataset Product H0133 </Description>
     <resContext>NASA.PDS</resContext>
     <resAggregation>null</resAggregation>
     <resClass>data.granule</resClass>
     <resLocation>iiop://PDS.ProfServer.GEO.ODY.GRS</resLocation>
     </resAttributes>
   <profElement>
     <elemName>FILE_SPECIFICATION_NAME</elemName>
     <elemValue>/ody_2001/xxx/H0133.DAT</elemValue>
   </profElement>
   <profElement>
     <elemName>INSTRUMENT_ID</elemName>
     <elemValue>HEND</elemValue>
   </profElement>
</profile>
```

Figure 8: An Example Profile for PDS

Message-based communications reduces the number of point-to-point connections as new data nodes are added, allowing the system to easily scale. Most importantly, PDS subsystems will remain geographically distributed and locally managed.

The first implementation of the information systems architecture, called PDS-D (Distribution) D01, supports on-line distribution of Mars Odyssey data products from six nodes. The second release, will include all Mars mission data and a third release following within a year will include the entire archive. Figure 9 illustrates the distributed systems architecture overlaid on the distributed PDS nodes.

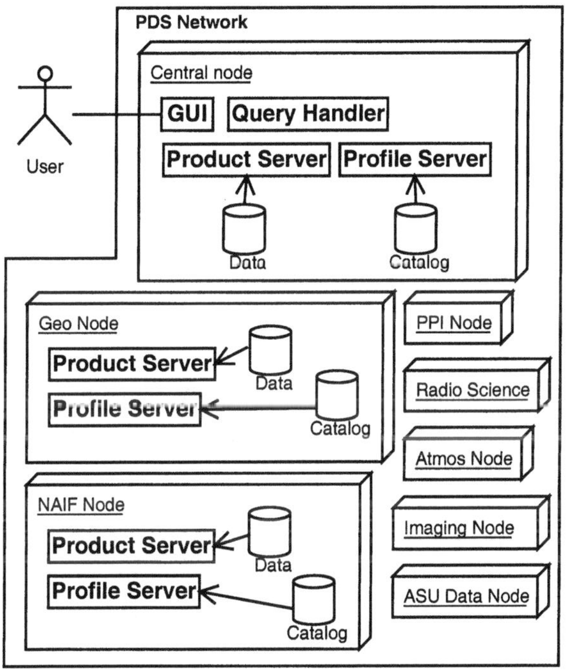

Figure 9: Example Distributed Component Architecture for the PDS-D D01

A representative example product that can be produced using a fully interoperable system is the Mars Global Surveyor (MGS) data coverage plot. In this plot, indicators showing the geographical location of MGS imaging, altimeter, and thermal emission spectrometer data products are overlaid onto a Mars image mosaic from the Viking Mars image data set. Four data sets at several geographically distributed data systems are required to be accessed in order to produce the plot.

As the number and volume of planetary missions and data sets produced by those missions increase, the opportunity to make new scientific discoveries by correlating data sets is driving the new paradigm to federate together the planetary information system community. The PDS is paving the way by delivering a new system that provides the basic infrastructure to create not only an active archive for NASA's deep space robotic missions, but the capability and standards to tie together this community's information resources.

4.2 Early Detection Research Network

In September 2000, the Jet Propulsion Laboratory and the National Institutes of Health signed an interagency agreement to research metadata-driven data architectures that allow for data sharing between biomedical researchers across widely distributed, diverse science databases. It was based on JPL's experience in building data architectures for space science. The initiative involves the development of a "prototype data architecture for the discovery and validation of disease biomarkers within a biomedical research network." [3] Within biomedical research, many systems are implemented to local standards, with limited collaboration between studies. The JPL/NIH agreement is to develop domain data architectures for communities that allow for sharing within specific biomedical collaborations.

NIH and JPL partnered with the National Cancer Institute's Early Detection Research Network (EDRN) to create an informatics data architecture to enable data sharing among cancer research centers. EDRN is a research network composed of over 30 institutions focused on advancing translational research of molecular, genetic and other biomarkers in human cancer detection and risk assessment [3]. One of the principal informatics goals of EDRN is to create a knowledge environment allowing data captured through scientific research studies to be shared by interconnecting each of the centers. A standardized data architecture would be developed that allows a distributed systems infrastructure to exchange information that can be understood by all nodes interconnected through this architecture.

The EDRN data architecture focused on the standardization of common data elements (CDEs). The intent of the CDEs is to provide the basis by which EDRN investigators capture and communicate information with other investigators. The CDEs focused on standard definitions for epidemiological and biospecimen data based on the ISO/IEC 1179 specification. Many of the participating institutions captured biospecimen data in local database systems independently implemented by the research and clinical informatics staff located at those institutions. This resulted in many different data models describing similar data sets. The Data Management and Coordinating

Center (DMCC) lead an effort to establish a set of common data elements along with a published data dictionary for EDRN. This created the basis for a data architecture and a language by which disparate EDRN data nodes could communicate.

Once the EDRN CDEs were defined, JPL and EDRN then adopted the Object Oriented Data Technology (OODT) framework. Profiles were created representing EDRN data resources located institutions. For example, a profile of an institution containing blood specimens for females between the ages of 40 and 70 would refer to a particular institution's data node. The profiles created a mapping between the EDRN data architecture and the physical implementation of that distributed system. This allowed queries for certain data types to be routed to the appropriate product server located at the institution that managed that data resource. The EDRN CDEs were used to create the variable data element portion of the profile. These allowed for the specific definition of data resources for the EDRN name space. See figure 10 below for an example of the profile.

```
<?xml version="1.0"?>
<!DOCTYPE profiles SYSTEM "http://somehost.jpl.nasa.gov/dtd/prof.dtd">
<profiles> <profile>
<profAttributes>
     <profId>1.1.1.1.102</profId>
     <profType>profile</profType>
     <profStatusId>ACTIVE </profStatusId>
</profAttributes>
<resAttributes>
     <Identifier>EDRN_NODE_1_PRODUCT_SERVER/Identifier>
     <Title>Node 1 Cancer Center Product Server</Title>
     <Format>text/html</Format>
     <Description>A product server providing access to a database of clinical
sample information for the Early Detection Research Network
(EDRN) at Node 1 Cancer Center.
     </Description>
     <Language>en</Language>
     <resContext>NIH.NCI.EDRN</resContext>
     <resAggregation>granule</resAggregation>
     <resClass>system.productServer</resClass>
     <resLocation>iiop://NIH.NCI.EDRN.NODE_1.PRODUCT_SERVER
     </resLocation>
</resAttributes>
<profElement>
     <elemId>STUDY_PARTICIPANT_ID</elemId>
     <elemName>STUDY_PARTICIPANT_ID</elemName>
     <elemDesc>EDRN Participant ID</elemDesc>
     <elemType>CHARACTER</elemType>
     <elemEnumFlag>F</elemEnumFlag>
</profElement>
</profile>
```

Figure 10: An Example Profile for EDRN

In addition to the construction of both the EDRN common data elements and the data profiles, the OODT query and profile services were deployed at the EDRN Data Management and Coordinating Center (DMCC). The profile server would serve profiles of available EDRN resources, and the query service would coordinate queries to disparate product servers. The role of the DMCC therefore would be to coordinate informatics activities across the EDRN providing both a programmatic and technical hub for the construction of a data sharing architecture for EDRN as shown in figure 11 below.

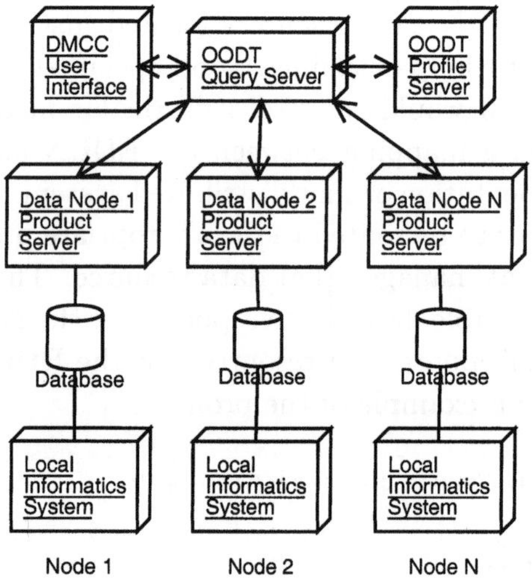

Figure 11: Example Distributed Component System Architecture for EDRN

Each participating data node created a data mapping between their local data model and data dictionary, and the EDRN data model and CDEs. This data mapping was used for both the construction of the profile as well as the construction of a node-specific product server [3.4]. The product server provided the translation necessary to query the data node and return the results using the EDRN CDEs and OODT messaging infrastructure. This allowed data nodes to retain their current database implementation, but be plugged into the EDRN informatics system.

One of the principal goals of the EDRN system was to standardize the interfaces to EDRN data resources, and abstract scientists away from having to understand the specific implementation of each system. This included a user interface developed by the DMCC to provide uniform queries across all data nodes, and to cluster results. The user interface submitted queries through the OODT system using the EDRN CDEs, received aggregated results from all sites mapped to the CDEs, and then presented the results in a unified form. While details of specific data nodes was also returned, users were unaware that the results were derived from federated databases located across North America.

Many of the challenges encountered involved semantic differences between the data captured at each of the sites. While the data was similar, the representation and information captured was not always orthogonal to the EDRN CDEs. For example, epidemiological data captured by one institution did not necessary contain the detail to map to the specific data values within the EDRN CDEs. Generalizations had to occur which often meant that queries for specific data values would not identify certain data resources that did not have sufficient granularity in their model to resolve the detailed query.

Finally, sharing of human subjects research data also presents challenges in terms of data security both from a technical and policy perspective. While the sharing of human subjects data presented different challenges and requirements to provide a tight security implementation due to the sensitive nature of the data, it was recognized that information retrieval of planetary and biomedical research data products could leverage a common systems approach. Both implementations leveraged a data architecture focused on using metadata as a common language for describing information products. This allowed the implementation of OODT to provide the data system framework for sharing information. In addition, it was concluded that clustering data from multiple nodes increased the ability for scientists to access and analyze data across multiple data systems without having to understand the physical location or topology of how the data was defined across multiple disparate data sources.

5 Deployment and Maintainance

Deploying the first OODT system into production proved challenging. The first deployment was the National Cancer Institute's Early Detection Research Network project (see Section 4.2 for more details). This deployment was instructive and resulted in a revised and highly successful deployment that took advantage of the lessons learned from the first.

5.1 Requirements

JPL faced three specific problems in deploying and debugging a distributed information clustering system:

- Debugging and patching framework and query handler software at remote sites.
- Reconfiguring and debugging CORBA (and CORBA over SSL) communications between sites.
- Maintaining component availability at each site.

Although remote logins to Unix systems would have faciliated meeting these requirements, many of the EDRN sites were not running Unix. Moreover, there were security concerns in running remote desktop applications.

Instead, it was decided to include remote debugging and server process management software as part of the OODT software itself. By integrating it with the OODT framework, the administrators running the OODT software would provide a way for JPL to debug information clustering programs and patch software and configuration files. The system would also automatically restart downed server processes with configurable criteria and save events for analysis from JPL or anywhere in the world.

To install the Server Manager, we created a Windows installer typical for Windows programs. By making the Server Manager as easy to install as possible, it was faciliated for administrators to have their sites added to the EDRN with a minimal set of components. Later configuration could occur from JPL. (Unix sites face a more traditional Unix application installation.)

5.2 Client/Server Architecture

The Server Manager is a server process that in turn manages other server processes on a single host system. Client applications run by the developers at JPL communicate with each remote Server Manager in order to manipulate those managed server processes.

The communications transport is HTTP, generally regarded as safe and effective. And XML-RPC provides simple yet useful semantics over HTTP. Also useful is the fact that simple XML-RPC calls may be crafted by hand. Figure 12 shows a sample XML-RPC call that starts a process named "myServer."

```
<methodCall>
  <methodName>serverMgr.proc.myServer.start</methodName>
</methodCall>
```

Figure 12: Sample XML-RPC invocation

5.3 Responsibilites of the Server Manager

The Server Manager's primary job is to manage the processes that comprise the OODT information clustering framework. In general, however, it can manage any kind of server process, and is extensible to support additional kinds. The secondary responsibilities of the Server Manager are to allow remote manipulation of server processes: defining, creating, stopping, starting, and diagnosing of processes. Additionally, it allows access (with only permissions that it itself has) to the filesystem of the system on which it runs so that we can patch and upgrade software installations.

5.3.1 Process Management

The process lifecycle (see Figure 13) begins when a developer at JPL uses a management client to connect to a Server Manager at a remote site and defines a new process.

Various types of processes are managed, and each one uses specific criteria to test for liveness. For example, a web server is alive if a configurable test URL can be used to successfully retrieve a document from the server. A CORBA name server is alive if its ORB is able to receive and process method calls.

Internally, the Server Manager implements the process types as a singly rooted class hierarchy. Through subclassing, we can define new process types for specific needs without impacting the framework. Figure 14 shows the current conceptual class hierarchy.

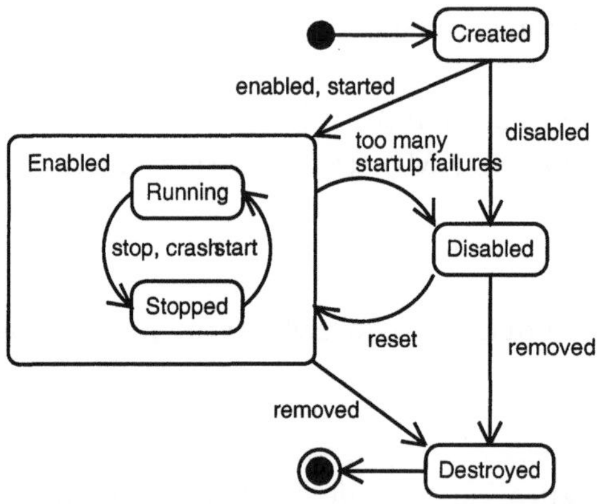

Figure 13: UML state diagram of the lifecycle running under a Server Manager

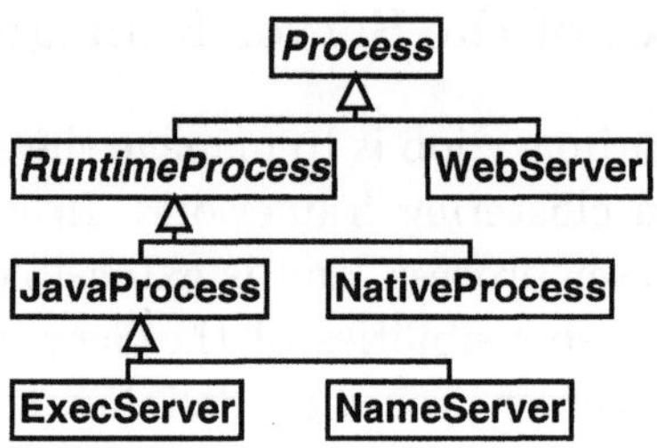

Figure 14: UML diagram of process class hierarchy

5.3.2 Automatic Restart

In order to provide as much availability of service without involving a JPL developer or other administrator, the Server Manager restarts processes configured for automatic restart. Each process has several configurable parameters that control how often a process gets restarted and preventing a badly behaved process from restarting too quickly. This helps maximize the "uptime" of information clustering applications.

If a process starts and stays running past its minimum healthy run time, the Server Manager considers it healthy and active. Any crash after its minimum healthy run time does not penalize the process. Crashes before such a time though incur a penalty. If this penalty exceeds its restart limit, the process is disabled. A disabled process isn't automatically restarted until an administrator manually restarts it or until its reset time expires. When the reset time expires, a disabled process is reenabled, its penalty slate wiped clean, and it is started again, if possible.

5.3.3 Remote Debugging

The Server Manager arranges to capture the output of every program it runs and makes that output available on demand. This feature enables debugging of remote processes.

Using the graphic management application, a developer can retrieve buffered output from a process for review. Typically, a developer may then patch the program, upload it using the Server Manager, restart the process, and retrieve the output again, repeating until the bug is resolved. Using this technique, JPL developers successfully created information clustering servers across the country for EDRN and are using the same system for the Planetary Data System application (see Section 4.1 for more details).

5.3.4 Patching Programs

After determining that a bug exists at a remote location, developers modify their local copies of the software and recompile. Server Managers at remote sites enable developers to install patched versions of software from their desktops. "Programs" in this sense includes both Java archive files as well as configuration files and other filesystem artifacts that implement the OODT information clustering framework.

Using the graphic management client, developers and administrators can connect to a Server Manager and install new files, download files, create directories, rename files and directories, delete files and directories, and set various file attributes.

5.3.5 Executing Scripts

By being able to execute scripts on the Server Manager, it's possible to provide features that aren't directly provided by the Server Manager itself. As an example, if a server process "detaches" itself from a Server Manager, it becomes impossible to control through the Server Manager. In this case, the developer writes a small script to locate and terminate the process and sends it to the Server Manager for execution at the remote site.

Because within the OODT team we had standardized on the Ant build environment, Ant was the natural choice for scripting language. Although intended as a configuration management and build tool, Ant has enough features to be considered a general scripting language. And since Ant is implemented in Java, adding it to the Server Manager took almost no effort.

6 Conclusions

A key to clustering information is creating the architecture that allows for the organization of the data. The OODT architecture enables the separation of the data and technology architecture layers. This allows for the evolution of these two architectural approaches to occur independently. By focusing on generic methods for describing information resources, OODT is able to provide a software framework for information retrieval across many different science domains.

Our experience has shown that development of the data architecture can be very difficult. It requires pairing both scientists or domain experts, and computer scientists together in order to develop the domain models which describe the data architecture. Creating the model allows for interoperability if disparate data sources can provide the mapping between locally implemented data systems and the domain model.

The Planetary Data System's (PDS's) long history of metadata usage simplified deployment of a distributed services framework like OODT. Although still in its infancy, the deployment continues successfully. The PDS data architecture enables the clustering of information by simply connecting more and more data nodes. The single point of entry provides a unified view across disparate databases that make up multiple datasets, relieving the researcher from having to understand each site's protocols and information models.

Implementation of the EDRN application was possible through careful analysis of each participating site's data architectural models, and creation of a common model for the EDRN. Teams of biomedical and computer science researchers from several institutions constructed both the data and technology architecture using OODT. Development of the Common Data Elements (CDEs) was essential to providing a compatible vocabulary to work between sites. Software to enable the distribution and validation of the system was critical to deploying the infrastructure at geographically distributed locations. Like the PDS, a single web interface provides a unified view of what in reality are multiple differing databases containing multiple differing interpretations of cancer biomarker data.

Virtual clustering of scientific information, leveraging metadata and implemented through distributed services makes automatic correlation and unified views like these presented possible. While science domains differ, the technical approach to implementing distributed clustering and information retrieval systems can follow a very similar development path. As our research continues, we look forward to future deployments of the OODT framework in other applications, continuous refinements to capturing and expressing metadata and semantic models, and new technical architectures which can define common approaches to inter-relating cross-disciplinary information resources.

References

[1] Tim Bray et al. *Extensible Markup Language (XML) 1.0 (Second Edition)*, (Cambridge, World Wide Web Consortium, 2000).

[2] R. J. Hanisch. "ISAIA: Interoperable Systems for Archival Information Access," Final Report-NAG5-8629, (NASA Applied Information Systems Research Program, 2002).

[3] Daniel Crichton, Gregory Downing, J. Steven Hughes, Heather Kincaid and Sudhir Srivastava. "An Interoperable Data Architecture for Data Exchange in a Biomedical Research Network," (Bethesda, 14th IEEE Symposium on Computer-Based Medical Systems, 2001).

[4] Daniel Crichton, J. Steven Hughes, Jason Hyon and Sean Kelly. "Science Search and Retrieval using XML," (Washington, 2nd National Conference of Scientific and Technical Data, 2000).

[5] Dublin Core Metadata Initiative. *The Dublin Core Element Set Version 1.1*, (Dublin: DMCI, 1999).

[6] Li Gong. *Project JXTA: A Technology Overview*, (Palo Alto, Sun Microsystems, 2001).

[7] Erich Gamma et al. *Design Patterns: Elements of Reusable Object-Oriented Software*, (Reading, Addison-Wesley, 1995).

[8] *ISO/IEC 11179 - Specification and Standardization of Data Elements, Part 1-6*, (ISO/IEC specification, http://www.iso.ch/iso).

[9] Ned Freed and Nathaniel Borenstein. "Multipurpose Internet Mail Extensions (MIME) Part One: Format of Internet Message Bodies (RFC2045)," (Reston, The Internet Society, 1996).

[10] Websters New Collegiate Dictionary. (G.&C. Merriam Company, 1974).

[11] J.C. French, A. K. Jones, J. L. Pflatz. "A Summary of the NSF Scientific Database Workshop," (Quarterly Bulletin of IEEE Computer Society technical committee on Data Engineering, Volume 13, No. 3, September 1990).

[12] *Planetary Data System Standards Reference*, (JPL D-7669 Part 2, http://pds.jpl.nasa.gov/stdref).

Clustering and Information Retrieval (pp. 299-329)
W. Wu, H. Xiong and S. Shekhar(Eds.)

Granular Computing for the Design of Information Retrieval Support Systems

Y.Y. Yao
Department of Computer Science
University of Regina
Regina, Saskatchewan, Canada S4S 0A2
E-mail: yyao@cs.uregina.ca

Contents

1 Introduction

Information Retrieval (IR) systems were traditionally used in libraries as a tool for searching for relevant information with respect to the user information needs [39]. An IR system is designed with the objective to provide useful and only useful documents from a large document collection [3, 39, 45]. The introduction of the World Wide Web (the Web), digital libraries, as well as many markup languages, has offered new opportunities and challenges to information retrieval researchers [3].

The Web is a totally new media for communication, which goes far beyond other communication medias, such as paper form publishing, radio, telephone and television. It revolutionizes the way in which information is gathered, stored, processed, presented, shared, and used. In this study, we concentrate on one particular use of the Web as a media and tool supporting scientific research. The Web has significant impacts on academic research. In contrast to traditional libraries and paper form publishing, it is a new platform for carrying out scientific research [28]. The amount of scientific information, such as online journals, books and scientific databases, increases in a very fast speed [21]. The existence of many effective tools, such as search engines and online reference services, makes scientific literature immediately accessible to a large group of scientists [21]. Studies have shown that articles available on the Web are more highly cited and used [20, 28]. The Web becomes a large and searchable virtual library. The problem of making effective use of the Web for research is a challenge for every scientist.

A basic tool to support research through the Web is search engines. The

Web search engines are the most heavily-used online services [8]. The effective use of the Web depends on, to a large extent, the success of many search engines [21]. Many Web search engines are designed based on the principle of information retrieval [22]. They inherit many of the disadvantages of traditional IR systems. IR systems focus mainly on the retrieval functionality, namely, the selection of a subset of documents from a large collection. There is little support for other activities of scientific research. IR systems use simple document and query representation schemes. A document is typically represented as a list of keywords, and a query is represented as either a list of keywords or a Boolean expression. There is little consideration of the relationships between different documents and between different portions of the same document. Semantic and structure information in each document is not used. IR systems use simple pattern based matching method to identify relevant documents. The philosophy and technologies of IR may be sufficient to support scientific research in the conventional library environment, where structure and semantic information of documents is not readily available. IR systems are inadequate to support research on the new Web platform.

Many researchers have attempted to extend and modify IR systems to meet the new challenges brought by the Web [3]. For example, research has been done on the use of hyper-text documents [3], structured documents [2], semantic information represented by ontology [15], and automatic citation analysis [23]. The research by Lawrence's group covers many important topics and directions on the use of the Web for supporting scientific research (http://www.neci.nec.com/~lawrence/papers.html). However, many such existing studies focus on particular supporting functionalities in isolation. One needs to address many interrelated and interactive functionalities in a more general framework.

In this chapter, we introduce the notion of Information Retrieval Support Systems (IRSS) as a general framework for supporting scientific research [56, 58]. IRSS is viewed as the next generation in the evolution of retrieval systems. IRSS is based on a new design philosophy which emphasizes many supporting functionalities, in addition to the simple retrieval functionality. As a more concrete example to demonstrate the potential value of IRSS, we describe a Granular Computing (GrC) model for the organization and retrieval of scientific XML documents [58].

The rest of the chapter is organized as follows. Section 2 reviews basic concepts of granular computing. Section 3 reviews basic concepts of information, with emphasis on granulation of document space, query (user) space, term space, and retrieval results. In Section 4, we argue that IRSS is the next generation in the evolution of retrieval systems. Section 5 discusses the basic issues of IRSS. Section 6 describes in detail the organization and retrieval of scientific XML documents.

The main objective of this chapter is to draw attention of information retrieval researchers to IRSS. It is hoped that many related, but isolated studies, topics, techniques, tools, and systems can be unified under the umbrella of IRSS.

This chapter summarizes and extends our preliminary studies on IRSS, and draws many results from two recent papers [56, 58].

2 Granular Computing

As a recently renewed research topic, granular computing (GrC) is an umbrella term to cover any theories, methodologies, techniques, and tools that make use of granules (i.e., subsets of a universe) in problem solving [25, 54, 55, 61]. Basic ingredients of granular computing are subsets, classes, and clusters of a universe. They have been considered either explicitly or implicitly in many fields, such as data and cluster analysis, database and information retrieval, concept formation, machine learning, and data mining [53, 61].

2.1 Basic Issues

There are many fundamental issues in granular computing, such as granulation of the universe, description of granules, relationships between granules, and computing with granules. Issues of granular computing may be studied from two related aspects, the construction of granules and computing with granules. The former deals with the formation, representation, and interpretation of granules, while the latter deals with the utilization of granules in problem solving.

Granulation of a universe involves the decomposition of the universe into parts, or the grouping of individual elements into classes, based on available information and knowledge. Elements in a granule are drawn together by indistinguishability, similarity, proximity or functionality [61]. The interpretation of granules focuses on the semantic side of granule construction. It addresses the question of why two objects are put into the same granule. It is necessary to study criteria for deciding if two elements should be put into the same granule, based on available information. One must provide necessary semantic interpretations for notions such as indistinguishability, similarity, and proximity. It is also necessary to study granulation structures derivable from various granulations of the universe [59]. The formation and representation of granules deal with algorithmic issues of granule construction. They address the problem of how to put two objects into the same granule. Algorithms need to be developed for constructing granules efficiently.

Computing with granules can be similarly studied from both the semantic and algorithmic perspectives. On the one hand, one needs to interpret various relationships between granules, such as closeness, dependency, and association, and to define and interpret operations on granules. On the other hand, one needs to design techniques and tools for computing with granules, such as approximation, reasoning, and inference.

According to Zadeh [61], granular computing suggests the basic guiding principle of fuzzy logic:

> "*Exploit the tolerance for imprecision, uncertainty and partial truth to achieve tractability, robustness, low solution cost and better rapport with reality.*"

It offers a more practical philosophy for real world problem solving. Instead of searching for the optimal solution, one may search for good approximate solutions. One only needs to examine the problem at a finer granulation level with more detailed information when there is a need or benefit for doing so.

It should be pointed out that studies of granular computing are only complementary to vigorous investigations on precise and non-granular computational approaches. The latter may provide justifications and guidelines for the former.

2.2 Simple Granulations

Let U be a finite and non-empty set called the universe, and let $E \subseteq U \times U$ denote an equivalence relation on U. The pair $apr = (U, E)$ is called an approximation space [29]. The equivalence relation E partitions the set U into disjoint subsets. This partition of the universe is called the quotient set induced by E and is denoted by $U/E = \{[x]_E \mid x \in U\}$, where

$$[x]_E = \{y \mid y \in U, xEy\}, \tag{1}$$

is the equivalence class containing x. The equivalence relation is the available information or knowledge about the objects under consideration. It represents a very special type of similarity between elements of the universe. If two elements $x, y \in U$ belong to the same equivalence class, we say that x and y are indistinguishable, i.e., they are similar. Each equivalence class may be viewed as a granule consisting of indistinguishable elements. It is also referred to as an equivalence granule. The granulation structure induced by an equivalence relation is a partition of the universe. There is a one-to-one correspondence between equivalence relations and partitions of the universe.

A partition is only a very restricted granulation of the universe, in which no overlap between granules is allowed. In general, one can use a covering of the universe to granulate a universe. In this case, a universe is divided into a family of possibly overlap granules. By allowing the overlap between granules, one can put an element into more than one granule.

There is no longer a one-to-one correspondence between coverings of a universe and certain type of binary relations on the universe. Nevertheless, one may still construct a covering using some specific type of binary relations. Suppose that a binary relation $R \subseteq U \times U$ is used to represent the similarity between elements of the universe. It is reasonable to assume that similarity is at least reflexive, but not necessarily symmetric and transitive [41]. For a reflexive binary relation R on U, if xRy, we say that y is R-related to x. A binary relation may be more conveniently represented using successor neighborhoods, or successor granules [53]:

$$(x)_R = \{y \mid y \in U, xRy\}. \tag{2}$$

The successor neighborhood $(x)_R$ consists of all R-related elements of x. When R is an equivalence relation, $(x)_R$ is the equivalence class containing x. When R is a reflexive relation, the family of successor neighborhoods $U/R = \{(x)_R \mid x \in U\}$ is a covering of the universe, namely, $\bigcup_{x \in U}(x)_R = U$.

A partition or a covering can be viewed as a brief description, or coarsened and granulated view, of the universe. We consider each equivalence class in a partition or a class in a covering as a whole instead of many individuals. In general, the number of the equivalence classes or subsets in a covering are much smaller than the number of the elements of the universe. The coarsening of the universe enables us to ignore or omit some fine differences between distinct objects. In other words, the coarsened view may allow us to observe some basic structures of the universe, which may not be easily observable under the very fine description of objects. The coarsening of the universe may also reduce computational costs of operations on the universe.

2.3 Hierarchical Granulations

The simple one-level granulated views of a universe are based on binary relations representing the simplest type of similarities between elements of the universe. Two elements are either related or unrelated. To avoid such a limitation, multi-level granulation structures can be constructed by putting together simple granulation structures [55]. Each level of the complex structure is a simple granulation structure such as a partition or a covering.

A multi-level hierarchical granulation structure can be interpreted and constructed by a nested sequence of binary relations [26, 30, 55]. Recall that a binary relation on U is a subset of the Cartesian product $U \times U$. The set inclusion defines an order on binary relations on U. An equivalence relation E_1 is said to be finer than another equivalence relation E_2, or E_2 is coarser than E_1, if $E_1 \subseteq E_2$. For the corresponding partitions, we write $U/E_1 \preceq U/E_2$. A finer relation produces smaller granules than a coarser relation, i.e., $[x]_{E_1} \subseteq [x]_{E_2}$ for all $x \in U$. Each equivalence granule of E_2 is in fact a union of some equivalence granules of E_1. Each granule of E_1 is obtained by further partitioning a granule of E_2. In general, we may consider a nested sequence of m equivalence relations:

$$E_1 \subseteq E_2 \subseteq \ldots \subseteq E_m. \tag{3}$$

The corresponding sequence of equivalence granules satisfies the condition:

$$[x]_{E_1} \subseteq [x]_{E_2} \subseteq \cdots \subseteq [x]_{E_m}. \tag{4}$$

The nested sequence of equivalence relations produces a multi-level partitions of the universe. This leads to a simple multi-level granulation structure of the universe. Different granulations of the universe form a linear order. A partition is either a refinement or a coarsening of another, although some granules in different levels may be the same.

Let $U = \{a, b, c, d, e, f\}$ be a universe. An example of hierarchical granulation defined by a nested sequence of equivalence relations is given by:

$$\begin{aligned}
\pi_5 &: \ \{\{a, b, c, d, e, f\}\}, \\
\pi_4 &: \ \{\{a, b, c, d\}, \{e, f\}\}, \\
\pi_3 &: \ \{\{a, b\}, \{c, d\}, \{e, f\}\}, \\
\pi_2 &: \ \{\{a\}, \{b\}, \{c, d\}, \{e, f\}\}, \\
\pi_1 &: \ \{\{a\}, \{b\}, \{c\}, \{d\}, \{e\}, \{f\}\}.
\end{aligned}$$

The linear order on the partitions is given by $\pi_1 \preceq \pi_2 \preceq \pi_3 \preceq \pi_4 \preceq \pi_5$. Partition π_5 is the coarsest partition, and π_1 is the finest partition. An equivalence class in a coarser partition is further divided into smaller equivalence classes in a finer partition.

Another type of multi-level granulation structures can be defined by a nested sequence of coverings or a nested sequence of reflexive binary relations. Each level is a covering of the universe. For two reflexive binary relations R_1 and R_2, U/R_1 is called a finer covering than U/R_2, written as $U/R_1 \preceq U/R_2$, if $R_1 \subseteq R_2$. For a pair of arbitrary coverings τ_1 and τ_2, we say that τ_1 is a finer covering than τ_2, written as $\tau_1 \preceq \tau_2$, if every class in τ_1 is a subset of certain class in τ_2. Suppose we have a sequence of reflexive binary relations:

$$R_1 \subseteq R_2 \subseteq \ldots \subseteq R_m. \tag{5}$$

Each relation defines a covering of the universe U/R_i and together they define multi-level coverings.

An example of multi-level coverings is given by:

$$\begin{aligned}
\pi_5 &: \{\{a,b,c,d,e,f\}\},\\
\tau_4 &: \{\{a,b,c,d\},\{a,b,c,f\},\{a,f\}\},\\
\tau_3 &: \{\{a,b,c\},\{a,d,e\},\{a,e,f\},\{a,f\},\{b,c\}\},\\
\tau_2 &: \{\{a,b\},\{a,c\},\{a,d\},\{b\},\{e,f\},\{f\}\},\\
\tau_1 &: \{\{a\},\{b\},\{c\},\{d\},\{e\},\{f\}\}.
\end{aligned}$$

A class in a coarser covering is obtained by adding some elements to a class in a finer covering. In other words, more elements are considered to be similar in a coarser covering.

In a hierarchy, one typically associates a name with a cluster such that elements of the cluster are instances of the named category or concept [16]. Partitions (coverings) in higher levels may be viewed as generalization of partitions (coverings) in lower levels, while partitions (coverings) in lower levels as specialization of partitions (coverings) in higher levels. A name given to a cluster in a higher level is more general than a name given to a cluster in a lower level, while the latter is more specific than the former.

3 Information Retrieval

The discipline of information retrieval concerns the organization, analysis, storage, searching, acquisition and dissemination of information. An information retrieval system is designed with the objective of identifying or ranking useful items from a large quantity of stored information in response to a user request. While information retrieval systems were traditional used in libraries, their most modern use is the Web search engines. Such systems are used extensively by scientists as an effective tool to find relevant information.

3.1 Basic Issues and Problems

Three fundamental issues in information retrieval are document representation, query formulation, and retrieval functions [3, 7, 39, 45, 50]. One needs to design and implement an appropriate scheme to represent the contents of documents, a language to express user queries, and a retrieval function to search for relevant documents. Index terms play the connecting role between documents and queries (users). A document is considered to be relevant to a query if the user submitting the query judges the document to be useful.

Different retrieval models have been developed, such as the Boolean, vector space, and probabilistic models [39, 45]. Although they provide us formal and elegant formulations of information retrieval problem, they suffer from several shortcomings. The classical retrieval models are over-simplification of the real world retrieval problem.

Each IR model represents a document simply as a list of (weighted) terms that appear in the document. The list is typically the results of some statistical analysis of document text. It should be realized that the effectiveness of statistical analysis, although providing us useful information, is limited [45]. To avoid this difficulty, one needs to consider the structure information and semantic information of the document. Two levels of structure information can be used. The document level structure information shows the connection between components of a document. Such information is now readily available from the online searchable and structured documents prepared using some markup languages. The collection level structure information shows the connections of documents. Citation and co-citation analysis and subject clustering of documents are examples that explore the collection level structure information. Natural language analysis tools and ontology can also be used to discover both levels of structure information. In many information retrieval systems, all documents are described in the same level of details. The same document representatives are used independent of individual users. The lack of consideration of the diversity, background and intentions of users effects the performance of IR systems. It is expected that multiple and personalized representations of documents may be more effective.

Similar observations can also be made regarding the issue of query formulation. Typically, a query language is used to express user information needs. A query language may be either too restrictive to be very effective, or too complicated to be practically useful. The problem is made worse when a user is not clear what is being searched for. Many search engines accept natural language queries. However, such queries are in many cases translated into a list of keywords or some simple Boolean expression. An effective information retrieval system should support many query languages and tools for expressing user information needs.

Many information retrieval systems use a simple and single retrieval method independent of users. In addition, retrieval is based on keyword level matching. Documents containing the keywords appearing in the query are retrieved or ranked higher. Other information that may suggest the relevance of documents is not fully explored. Recent studies on text mining, user behavior analysis, and agent technology may provide potential solutions to such a problem [57]. One needs also to explore the potential of multi-strategy retrieval exemplified by meta-search engines.

In summary, on the one hand, many shortcomings of traditional information retrieval systems and Web search engines make them inadequate to support scientific research using the Web. One the other hand, the advances made in related fields, such as text mining, intelligent agents, and markup languages open new doors to expand information retrieval systems. In fact, many different special purpose systems have been implemented. Two examples of such systems are DBLP (http://dblp.uni-trier.de/) and ResearchIndex (http://citeseer.nj.nec.com/cs). The emergence of such systems strongly suggest the needs for a more general framework for the study of information retrieval to support scientific research, especially on the new platform of the Web.

3.2 Document Space Granulations

Document clustering is a widely used technique in information retrieval to reduce computational costs and improve retrieval effectiveness [34, 49]. Documents may be clustered in several ways, such as content based, query based, and citation based approaches. The most commonly way of clustering is content or topic based. Documents with similar content or topic are put into the same cluster [39, 45]. The recent extensive studies and renewed interest on text categorization further explore document clustering [17, 52]. Another type of document clustering methods is the query oriented document clustering. Documents are clustered based on their joint relevance to a set of queries. That is, documents are put into the same cluster if they tend to be relevant at the same to some queries [38, 60]. A similar idea is to use citation and co-citation information for document clustering [11]. Such clustering methods are used in ResearchIndex, in which, for example, co-cited documents are put into a cluster. Other methods for clustering documents is based on special characteristics of documents. For example, documents can be clustered based on authors, journals or conference, as is done in DBLP. Those clustering methods are not only valuable for content oriented retrieval, but also suitable for other special purpose retrieval.

In content based document clustering, a collection of documents is divided into clusters such that each cluster consists of similar documents. A center called centroid is constructed for each cluster to represent all the documents in that cluster [39]. A clustering of documents provide a granulated view of the document collection. One may use either a partition or a covering of the document collection for clustering. A hierarchical clustering of documents is produced by decomposing large clusters into smaller ones. The large clusters offer a rough representation of the document. The representation becomes more precise as one moves towards the smaller clusters. A document is described by different representations at various levels. Hence, a cluster-based IR system implicitly employs multi-representation of documents. Cluster based retrieval is done by comparing a query with the centers of the larger clusters. If the center of the current cluster is sufficiently close to the query, then the query will be compared against the centroid of the smaller clusters at a lower level. In other words, if it is concluded that a document is not likely to be useful using a rough description, then the document will not be further examined using more precise descriptions. Different retrieval methods may also be employed at different levels.

Document clustering only reduces the dimensionality of the document collection while the dimensionality of index terms remains the same. That is, the same number of terms is used for the representation of cluster centers regardless of the level in the document hierarchy. On the other hand, text categorization uses some predefined categories to label or name a cluster, and thus introduces different representations of the same document.

3.3 Query (User) Space Granulations

Like the granulation of document space, one can construct granulated views of query (user) space in several ways, such as content based, document based approaches [46, 47]. Content based query clustering is similar to content based document clustering [38]. The similarity of queries is evaluated based on index terms used by the queries. Similar queries are grouped together to represent the needs of a group of users. Content based approaches can be easily extended to cluster users based on user profiles or user logs. On the other hand, document based query clustering methods use the overlap of relevant documents, retrieval results, of queries [10, 33]. Although two queries may not be similar according to their contents, they are still considered to be similar due to a large overlap of relevant documents.

Some recent ideas for query clustering are related to the document based query clustering [47]. They are useful in question answering systems or search engines, such as AskJeeves (http://www.askjeeves.com/). Beeferman and Berger [4] suggested to cluster queries by using click-through data, which is implicit relevance information provided by users. Wen et al. [46, 47] combined both content based and document based (through user document clicks) approaches for query clustering.

3.4 A Unified Probabilistic Model

The application of granulated views of both document space and query space is well illustrated in a unified probabilistic model proposed and studied by Robertson et al. [35].

In probabilistic information retrieval models, the relevance of documents to queries is modeled in probabilistic terms. There are four sub-models in the unified model. Maron and Kuhns' model (Model 1) is based on the granulation of query (user) space [27]. Queries are clustered in order to compute the probability of relevance of a document to a group of queries (users). In contrast, Robertson and Sparck Jones' model (Model 2) is based on the granulation of document space [36]. Documents are clustered in order to compute a probability of relevance of a group documents to a given query (user). With respect to the granulation of both document document and query spaces, one obtains a lower model (Model 0) which computes the probability of relevance of a group of documents to a group of queries (users). When both document and query spaces are not granulated, a high level model (Model 3) is obtained. The high level model represents the ideal situation where the relevance of individual documents to individual queries is used. However, it is difficult to work with Model 3 in practice, as we always have granulated views of either document or query space. Any document representation scheme or query language leads to granulated view of document or query space. Documents (queries) with the same description are considered to be a whole unit and can not be differentiated. A solution to this difficulty is the combination of both Model 1 and Model 2 [35]. More specifically, the relevance of a particular document to a particular query is estimated by the relevance of the document to a group of queries and the relevance of a group of documents to the query.

3.5 Term Space Granulations

The problem of term clustering and its application in information retrieval have been studied by many authors [5, 32, 37, 39, 40, 42]. In a term hierarchy, a cluster may be assigned new terms as labels of the cluster. The new terms are more general than each individual term in the cluster. In general, a multi-level coverings may be more suitable. A more specific term may be described by more than one general terms.

A term hierarchy serves as an effective tool to summarize knowledge about a specific domain. Many domain-specific term hierarchies or concept hierarchies have been used in the organization and retrieval of scientific literature. Examples of term hierarchies are the ACM Classification System (http://www.acm.org/class/) and the Mathematics Subject Classification (http://www.ams.org/msc/). With such systems, one immediately derives a hierarchical granulation of documents. At each level, documents are described by different terms of different specificities. Documents described by the same terms in the classification system are naturally put into the same cluster.

A main consideration in using term hierarchies is the trade-off relationship between the high dimensionality of index terms and the accuracy of document representation. One may expect a more accurate document representation by using more and specific index terms. However, the increase of the dimensionality of index terms also leads to a higher computational cost.

Term clustering techniques have been used mainly for retrieval, such as query modification, query expansion, and sophisticated retrieval functions [39]. Their use in the granulation of document space and query space has not been fully investigated [51]. The potential of a term (concept) hierarchy, especially its implied knowledge structure, needs to be fully exploited. The fast growing interest on ontology clearly demonstrates a trend in exploring relationship between terms (concepts).

3.6 Retrieval Results Granulations

In many situations, the list of documents returned by information retrieval systems such as search engines is too long and contains duplicate or near-duplicate documents [18]. In order to resolve those problems, many authors suggested and studied the granulation of retrieval results [1, 13, 19, 6, 43, 62]. By clustering retrieval results, one can organize the results and provide coarse-grained summarization to users.

The idea of granulating retrieval results was first studied by Preece [31, 43]. Willet [48] referred to such document clustering as query specific document classification, in contrast to the query independent document granulation discussed earlier. With respect to a particular query, clustering results are more effective [6, 13, 43].

An important issue in query specific document clustering is to obtain a meaningful description of the derived clusters to be presented to the user [19, 6]. The classical centroid based approach is no longer appropriate. It has been suggested that a few titles and some terms can be used as the description of a cluster [6, 13, 19]. One may also extract some important sentences from the documents in a cluster as a description of the cluster.

3.7 Structured and XML Documents

The granulated views of documents and terms can be exploited to discover collection level structures and organize documents accordingly. The document level structure information can be obtained from the use of markup language.

In information retrieval, full text and structured documents have been considered by many authors [2, 14]. With the development of XML (eXtensible Markup Language, http://www.w3.org/TR/REC-xml), there is a growing interest in the organization and retrieval of structured and semi-structured documents. XML is becoming a new standard for data representation and exchange. More XML documents are expected to be available on the Web.

In XML, the structures and the meaning of data are explicitly indicated by element tags. The structure of a document and element tags are defined through a DTD (Document Type Definitions). For example, a scientific article has a hierarchical (multi-level) granulated description given by the tree structure of DTD. An XML document can be viewed in many different ways by focusing on different tags. For example, one can easily extract only theorems from an XML document. More specifically, an XML document itself can be viewed as the physical view of the documents, and many different logical views can be obtained.

We can explore the rich information provided in XML documents. For example, one can cluster documents using certain tag fields. In this way, different granulated views of the collection can be formed with respect to different users. An XML document collection also allows multi-strategy retrieval. One may use structured queries by focusing on certain tags or perform free text retrieval by simply ignoring all tags. The rich information available in XML documents enable us to extend the traditional functionalities of information retrieval systems.

4 Evolution of Retrieval Systems

In order to extend information retrieval systems and search engines, we need to move beyond the simple retrieval and search centered design philosophy. To search for a new design philosophy of IRSS, we examine the evolution process of retrieval systems and the roles played by each type of retrieval systems [56]. It is concluded that IRSS is the next evolutionary stage of retrieval systems, which deals with more difficult and more complex problems, and provides more supporting functionalities to a user.

4.1 From Data Retrieval Systems (DRS) to Information Retrieval Systems (IRS)

In one of the classic books on information retrieval (IR), van Rijsbergen compared data retrieval (DR) and IR to illustrate the range of complexity of each mode of retrieval, as well as their common features and differences [45]. DRS may be considered as an early stage, and IRS as the next evolutionary stage in the development of retrieval systems.

In both modes of retrieval, on the one hand, there is a set of information items (data or documents), and on the other hand, there is a user information need. The function of a retrieval systems is to match items with the user need. Both DR and IR focus on the retrieval functionality, namely, the match of items and user information needs. It is not surprising that many results from the discipline of pattern recognition, which deals with pattern match, are applied into IR [9].

The differences between DR and IR can be seen from the ways in which information items and user information needs are represented, as well as the matching process [3, 45, 50]. A database system is a typical example of DR. In DR, data items and user information needs can be precisely described by using well understood knowledge representation schemes and query languages. The model is deterministic in the sense that the relationships between data items and user needs are well and objectively defined, and consequently exact match and deductive inference can be used. In contrast, in IR, documents and user needs can not be precisely described, and their relationships are ill and subjectively defined. The model is non-deterministic in the sense that partial or best match and inductive inference are used. In summary, DR deals with structured, and well-defined problems in which there is no uncertainty, while IR deals with semi-structured, unstructured, and ill-defined problems where uncertainty plays a major role.

In the design of IRS, the search functionality of DRS is extended. However, the design philosophy remains to be the same. Both DRS and IRS are designed to provide essentially the search functionality.

Some of the reasons for adopting the search functionality centered design philosophy can be seen from the historical roles played by IR systems. IR systems were first introduced in libraries for retrieval information [3]. Baeza-Yates and Ribeiro-Neto divide IR systems into three generations [3]. The first generation is basically an automation of conventional manual catalog search, which allows searches using author name and title. The second generation increases the search functionalities by allowing text based search, such as searching by subject headings, keywords, and more complex queries. The third generation focuses on improved graphical interfaces, hypertext features, and open system architectures. We can conclude that IR systems, in the context of libraries, only attempt to automate the task of search, while other user support functionalities are left to the librarians.

Baeza-Yates and Ribeiro-Neto differentiate two distinct types of user tasks when using a retrieval system, the retrieval (search) task and the browsing task [3]. A retrieval task is normally performed by translating an information need into a query and searching using the query. A browsing task is carried out by looking around in a collection of documents through an interactive interface. During browsing the user information need or objective may not be clearly defined, and can be revised through the interaction with the system. The third generation of IR systems provide more functionalities for browsing. As summarized by Baeza-Yates and Ribeiro-Neto, "Classic information retrieval systems normally allows information or data retrieval. Hypertext systems are usually tuned for providing quick browsing. Modern digital library and Web interfaces might attempt to combine these tasks to provide improved retrieval capabilities."

A careful examination of the three generations of IR systems shows that each later generation provides enhanced or more functionalities of the previous generation. As we move from DR to IR, as well as from one generation of IR systems to another, two dimensions of changes can be observed. One dimension concerns the complexity and the nature of the problems, ranging from structured, semi-structured, to unstructured. DR deals with structured problems, where all involved concepts can be precisely defined. IR deals with semi-structured problem, where some concepts can not be precisely defined. Future systems may deal with unstructured problems. Another dimension concerns the user control or user tasks, ranging from simple to complex. DR systems deal with fact retrieval, IR systems deal with non-fact searching and browsing. Future generations of systems may deal with more complex user tasks, such as analysis, organization, and discovery. This two-dimensional description is adopted from a well known framework for decision support systems (DSS) [44].

4.2 From Information Retrieval Systems (IRS) to Information Retrieval Support Systems (IRSS)

Many new developments in IR have been made, ranging from multimedia (images, audio or video) retrieval, hypertext retrieval, and digital library to Web information retrieval [3]. With the rapid development of the Web and digital libraries, we have witnessed a wider range of applications of IR systems, and a renewed interest in IR. In fact, IR systems such as search engines play an important role for the success and boom of the Web. On the other hand, the design philosophy and principles of IR, well discussed in classic textbooks [39, 45], have remained more or the less the same. The observation by Lesk that "the more things change, the more they stay the same" is still applicable today as it was applicable 10 years ago [24]. To a large extent, information retrieval can still be viewed as document retrieval by substituting 'document' for 'information', as pointed out by van Rijsbergen long time ago [45]. The fundamental ideas of IR systems, namely, indexing and searching, have remained to be the same. IRS are perceived as systems that provide the basic search and browsing functionalities.

From the two-dimensional description of DR and IR problems, we need to consider unstructured problem and more user control in the design and implementation of new information retrieval strategies and systems. We need to have a better understanding of user tasks. The ultimate goal of finding information is to use the relevant information, say, in a decision making process. For example, a researcher may compare, analyze and summarize the relevant information in preparing a scientific article, or evaluating a project proposal. In order to find and extract useful information from a large document collection or the Web, as well as effectively use the extracted information in problem solving, a user must play an active role in various tasks, such as browsing, investigating, analyzing, understanding, organizing, and searching the collection. Searching and browsing are only some of the simple and front end tasks. The next generation of IR systems must support more types of user tasks, in addition to searching and browsing. Since these tasks can not necessarily be described precisely, fully automation can not be expected. Instead, one can build various tools, methodologies, and languages for supporting such user tasks.

From the above discussion, we can conclude that the search centered philosophy for the design of IR system may no longer be suitable. This is also evident from the fact that many Web search engine users must spend more time to understand, filter, and organize documents returned by a search engine. With the Web as a new media for information storage, delivery, gathering, sharing, processing, and utilization, the problem of information retrieval is no longer a simple process of search.

A new set of philosophy and principles for the design and implementation of the next generation IR systems is needed. Instead of focusing on the search functionality, one focuses on the supporting of functionality [58]. This can be viewed as the next stage in the evolution of retrieval systems, which leads to the introduction of information retrieval support systems (IRSS). The objective of an IRSS is to support many different types of user tasks in finding and utilizing information, in a similar way that a decision support system (DSS) assists users in making managemental decisions [44].

5 Basic Issues of IRSS

In this study, the term information retrieval is used as a much broader umbrella term to cover any and every user tasks in finding and utilizing information from a collection of documents or the Web when performing scientific research. Similarly, the term IRSS is used as an umbrella term to describe any and every computerized systems used to support and improve information retrieval. In a more accurate sense, we perhaps should use the term Research Support System (RSS). We choose the term IRSS to emphasize the particular aspects of research support covered by conventional IR systems.

5.1 The Concept of IRSS

Our interpretation of IRSS draws extensively results from the related field of decision support systems (DSS) [44]. While DSS focus on supporting and improving decision making, IRSS focus on supporting and improving retrieval. The philosophy, principles, and techniques from DSS are applicable to IRSS by simply substituting the tasks of "decision making" for the tasks of "information retrieval". This view of IRSS is particular reasonable, if one considers the fact that it is necessary to obtain useful information in order to make intelligent and rational decision.

A classical definition of DSS given by Gorry and Morton [12] defines DSS as "interactive computer-based systems, which help decision makers utilize data and models to solve unstructured problems." Turban and Aronson compare different definitions of DSS and suggest the following one [44]:

> "Computer-based information systems that combine models and data in an attempt to solve nonstructured problems with extensive user involvement through a friendly user interface."

From the above definition, we want to stress two important features of DSS pertinent to our interpretation of IRSS. One feature is the combination of data and models. Data are raw and uninterpreted facts. In order to make sense of raw data, one needs to construct various models. Therefore, an DSS deals with both data and their interpretations. The other feature is the emphasis on the user involvement. An DSS plays a supporting role in problem solving.

One can give a formal definition of IRSS based on the definition of DSS. However, the definition of IRSS is not the focus of this chapter. It is perhaps wise to leave the notion loosely defined until we have gained more insights into the problem. For the time being, we rely on our intuitive understanding of IRSS.

5.2 Characteristics of IRSS

The problems of finding and using relevant information from a large collection of documents are unstructured problems that can not be easily and precisely described. It is made even more complicated by the fact that a user may not know exactly what is being searched for. In solving the retrieval problems, IRSS are more useful and effective than IRS. Based on the two features of DSS mentioned earlier, we can identify some characteristics that distinguish IRSS from current IRS, and show the advantage of IRSS.

Current IRS emphasis on the storage and search functionalities, which leads to a lack of consideration of the two important issues, namely, models and user involvement. In other words, an IRS performs search at the raw data level, instead of the model level, and without user interaction. Although recent IRS systems exemplified by Web search engines build hierarchical model to provide semantic interpretation of documents in a collection, the end users are not involved in the model construction process. A remedy to this problem is the use of personalized user profile and personalized bookmarks.

IRSS attempt to resolve the problems of IRS by providing more supporting functionalities. An IRSS provides models, languages, utilities, and tools to assist a user in investigating, analyzing, understanding, and organizing a document collection and search results. These tools allow the user to explore both semantic and structural information of each individual document, as well as the entire collection.

Three related types of models need to be considered in IRSS. Documents in a document collection serve as the raw data of IRSS. The document models deal with representations and interpretations of documents and the document collection. The retrieval models deals with the search. The presentation models deal with the representation and interpretations of results from the search. A single document model, a retrieval model, or presentation model may not be suitable for different types of users. Therefore, IRSS must support multi-model, and provide tools for users to manage various models.

The document models allow multi-representation of documents. Granular computing will play an important role in the construction of document models [51, 58, 61]. One can use a hierarchical granulation of document collection, namely, a layered and multi-resolution representation of documents. The same document is represented in less detail at a higher level than at a lower level. With the multi-model capability, a user can create different logical views of a document collection and logical links between documents, as well as compare and investigate various views. The resulting in-depth knowledge can help the user to locate and infer useful information. The recent development of XML enables us to describe both the structural and semantic information of a document. Such information makes the construction of multi-document models an achievable goal.

The retrieval models provide languages and tools to assist a user to performs tasks such as searching and browsing. IRSS should provide multi-strategy retrieval. A user can choose different retrieval models with respect to different document models. The presentation models allow a user to view and arrange search results, as well as various document models. The same results can be viewed in different ways by using distinct presentation models. Moreover, a user can analyze and compare results from different retrieval models.

IRSS implement and manage three types of models, as well as the associated languages, tools and utilities. An IRSS is highly interactive so that a user can make decisions at various stages. A user plays a more active role in the process of finding useful information. There are many advantages for the extensive user involvement. The user involvement is particularly important in exploratory type of searching and browsing. The usefulness or the relevance of each information item (i.e., a group of documents, a document, and parts of a document) can only be determined by the user. Without the involvement of a user in the analysis and organization of a document collection, one can not expect the user to provide a meaningful query.

From the previous discussion, we can conclude that an IRS performs easy and loosely structured retrieval tasks, where automation is possible and user involvement is not necessary. On the other hand, an IRSS supports a user to perform difficult and unstructured retrieval problems. Since a fully automation is impossible, at least not for the time being, the main function of IRSS is to support a user. The retrieval process is controlled by the user.

5.3 Components of IRSS

One may argue about the exact components of an IRSS. We take a simple approach by adopting the results from DSS and intelligent systems.

According to Turban and Aronson, a DSS normally consists of four subsystems: [44].

- *Data management subsystem:* This subsystem deals with lower level raw data management using software systems such as database management system (DBMS) and data warehouse.

- *Model management subsystem:* This subsystem is referred to as a model base management system (MBMS). It includes existing quantitative models for analyzing and interpreting the raw data, and provides language and tools for building user models.

- *Knowledge-based management subsystem:* This subsystem supports other subsystem and provides intelligence to a decision maker.
- *User interface subsystem:* This subsystem handles the interaction between user and the system.

The above schematic description of DSS can be applied to the study of IRSS. In other words, although the objects managed by each subsystem may be different, the fundamental principles are the same.

5.4 Fields Related to IRSS

Techniques, results and lessons from many fields can be used in the study of IRSS and to enhance the capabilities of an IRSS. A few related fields are summarized below, in addition to the previously discussed DSS and GrC.

Expert systems (ES).

A well established practice in expert systems is the separation of knowledge and inference engine. While the inference engine is logic based and problem independent, the knowledge base is domain specific. Expert system shells that implement inference engine can be used to build many different domain-specific expert systems. Similarly, we can build knowledge based IRSS by separating knowledge base and management subsystem. An IRSS shell can be built that provide a set of domain and user independent tools, using which domain specific IRSS can be constructed.

Users of retrieval systems may fall into many different categories, have different background, and with different types of information needs. Documents in a collection may also cover different domains. One can not expect to design a system that is best for everyone and for every domain. Consequently, one needs to study principles, methodologies, and techniques that can be used to design and implement domain-specific IRSS.

Another feature of expert systems is the explanation functionality. An expert system not only provides a solution, but can also explain why and how the solution is arrived. It is reasonable to insist on the explanation functionality of an IRSS.

Machine learning, data mining and text mining.

By applying algorithms of machine learning, data mining, and text mining to documents stored in an IRSS, one may discover patterns and extract knowledge useful to a user. Such functionalities are particularly useful to users who are interested in exploratory searching and browsing. For instance, a user can track trends in a particular area or discover emerging topics from the constantly changing document collection. A user may also discover links between different documents or research areas.

Computer graphics and data visualization.

In many cases, a user may not want details about particular documents that contain the useful information. A user may want to have a general feeling before going to a more in-depth analysis. With the granulation of document collection, it is possible to provide a user with granulated view, in which details are omitted. Most current IRS present search results in the form of ranked list of individual documents. In an IRSS, a user should be able to use graphics and visualization tools to view a particular document model. Visualization enables a user to perform high level inference and analysis.

Intelligent information agents.

Intelligent information agents have been used by many IRS to collect information and interact with users. The potentials of agents need to be further explored in IRSS. In particular, a user should be allowed to construct a personalized agent to interact with an IRSS. The autonomy and learning capabilities of agents make them attractive to both IRSS and users.

In summary, an IRSS can incorporate any particular type of information systems to provide a specific type of support.

6 A Granular Computing Model for Organizing and Retrieval XML Documents

In this section, we provide a concrete example to illustrate the ideas developed in the previous sections. We will focused mainly on the granulation of an XML document collections.

For simplicity, we consider scientific XML documents given by:

```
<Paper>
   <Title>  </Title>
   <Author>
      <Name>
        <First>  </First>
        <Last>   </Last>
```

```
      </Name>
      <Institution> </Institution>
   </Author>
   <Abstract>
      <Paragraph>  </Paragraph>
      ...
    </Abstract>
    <Section>
        <Title>  </Title>
        <Subsection>
          <Title>  </Title>
          <Paragraph>  </Paragraph>
          ...
        </Subsection>
        ...
    </Section>
    ...
</Paper>
```

Many tags, such as definition, theorem, etc., are not included, as the simple layout is sufficient for the purpose of illustration. It follows that the structure of an article can be clearly observed.

We have a natural multi-level granulation of a scientific document. The granulated representation of an article is the document level granulation. The semantic information provided by tags allows us build different logical views of the same document. For example, one can read only the section titles or the abstract to gain general knowledge about the article. One may also zoom in to particular paragraphs, definitions, or theorems for more details.

At higher levels of granulation, namely, the collection level granulation, we seek for relationships between individual XML documents. Such relationships can be derived from information in each document. In fact, a tag or a combination of tags may be used to cluster documents together to form granules. For example, the collection may be divided by the author name and each granule consists of articles published by that author. Each such granule may be further divided by the year of publication, so that we can obtain a more detailed picture of the publication record of that author. Similarly, documents can be grouped together by journal names, and further divided by the year of publication. Documents may be grouped together by topics, as represented by index terms (keywords) in the title, section titles, or the entire text, or citation. Those logical views of collection have also been used in some digital library systems, such as DBLP (http://dblp.uni-trier.de/) and ResearchIndex (http://citeseer.nj.nec.com/cs). However, each system has its predefined logical views, and a user is not allowed to create his/her own logical view. By providing the user with more flexibility, one can build a more flexible information retrieval support system.

With respect to the classification of document by keywords, we can build a hierarchical granulation structure. Since the title of an article is the coarsest representation, we can obtain the first level of granulation by using keywords in the title. It is expected that large sized granules would be produced. Each granule can in turn be further divided by keywords in the titles of sections and subsections. The resulting granules can be further divided by using keywords in the full text of the article. The article level granulation forms the remaining part of the hierarchy. At each level, the same document is represented differently. For example, at the collection level, it is represented as different lists of keywords, while at the article level, it is represented in XML format. The granulation of the collection enables the user to understand structural information about the collection. Such useful information may not be obtained from fixed and pre-determined logical views provided by some systems. In searching for useful information, the user can control the level of details through interaction with a retrieval support system. In the process of granulation, the user can enquiry the information of the documents in particular granules. Different retrieval strategies and methods may be employed. For example, at a higher level of the hierarchy, the user may choose a Boolean model to eliminate apparently uninteresting documents. At a lower level, the user may use the vector space model to obtain more accurate and ranked results.

In the GrC model, the three basic types of operations, creation of logical views (granulation), navigation through different logical views, and retrieval, support the user to investigate, to analyze, to understand the collection, and to search for useful information from the collection.

7 Conclusion

In this chapter, we discuss two related topics, namely, the application of granular computing to information retrieval and the introduction of Information Retrieval Support Systems (IRSS).

It is demonstrated that granulation plays an important role in information retrieval. Many entities in information retrieval can be granulated to obtain coarse-grained views. One can granulate document space, query (user) space, term space and retrieval results. The granulated views allow us to focus on the useful structures without looking into too much details. Many existing studies deal with the granulation of a specific entity. The interactions between granulations of different entities need further investigation.

Our formulation and understanding of IRSS draw extensively results from DSS, which in turn draw results from many fields. It represents a new design philosophy to information retrieval. It is hope that more attention can be drawn to this important topic. IRSS may be viewed as the next stage in the evolution of retrieval systems. IRSS focus on various supporting functionalities of retrieval systems, in addition to support search and browsing. IRSS also stress the importance of extensive user involvement.

The new philosophy of retrieval support, rather than retrieval, may fundamentally change the design goal of current IR systems, which enables us to move from simple retrieval systems to advanced retrieval support systems.

References

[1] R.B. Allen, P. Obry and M. Littman, An interface for navigating clustered document sets returned by queries, *Proceedings of the ACM Conference on Organizational Computing Systems* (1993) pp. 166-171.

[2] J. André, R. Furuta and V. Quint (eds.), *Structured Documents* (Cambridge, Cambridge University Press, 1989).

[3] R. Baeza-Yates and B. Ribeiro-Neto, *Modern Information Retrieval* (New York, Addison Wesley, 1999).

[4] D. Beeferman and A. Berger, Agglomerative clustering of a search engine query log, *Proceedings of KDD'00* (2000) 407-415.

[5] H. Chen, T. Ng, J. Martinez and B. Schatz, A concept space approach to addressing the vocabulary problem in scientific information retrieval: an experiment on the worm community system, *Journal of the American Society for Information Science* Vol. 48 (1997) pp. 17-31.

[6] C. de Loupy, P. Bellot, M. El-Bèze and P.F. Marteau, Query expansion and classification of retrieved documents, *Proceedings of the Seventh Text REtrieval Conference (TREC-7)* (1998) pp. 382-389.

[7] S. Dominich, *Mathematical Foundations of Information Retrieval* (Dordrecht, Kluwer Academic Publishers, 2001).

[8] M.P. Courtois, and M.W. Berry, Results ranking in Web search engines, *Online* Vol. 23 (1999) pp. 39-46.

[9] R.O. Duda and P.E. Hart, *Pattern Classification and Scene Analysis* (New York, Wiley, 1973).

[10] L. Fitzpatrick and M. Dent, Automatic feedback using past queries: social searching? *Proceedings of SIGIR'97* (1997) pp. 306-313.

[11] Garfield, E. *Citation Indexing – Its Theory and Application in Science, Technology and Humanities* (New York, John Wiley & Sons, 1979).

[12] G.A. Gorry and M.S.S. Morton, A framework for management information systems, *Soloan Management Review* Vol. 13 (1971) pp. 55-70.

[13] M.A. Hearst, J.O. Pedersen, Reexamining the cluster hypothesis: Scatter/Gather on retrieval results, *Proceedings of SIGIR'96* (1996) pp. 76-84

[14] F.C. Heeman, Granularity in structured documents, *Electronic Publishing* Vol. 5 (1992) pp. 143-155.

[15] J. Heflin, R. Volz, and J. Dale (eds.), *Requirements for a Web Ontology Language*, W3C Working Draft, 07 March 2002, http://www.w3.org/TR/webont-req/

[16] N. Jardine and R. Sibson, *Mathematical Taxonomy* (New York, Wiley, 1971).

[17] T. Joachims, Text categorization with support vector machines: learning with many relevant features, *Proceedings of the 10th European Conference on Machine Learning* (1998) pp. 137-142.

[18] J.W. Kirriemuir and P. Willet, Identification of duplicate and near-duplicate full-text records in database search-outputs using hierarchic cluster analysis, *Program* Vol. 29 (1995) pp. 241-256.

[19] Y. Kural, S. Robertson and S. Jones, Clustering information retrieval search outputs, *21st Annual BCS-IRSG Colloquium on IR* (1999), http://www1.bcs.org.uk/DocsRepository/02800/2856/kural.pdf

[20] S. Lawrence, Online or invisible, *Nature* Vol. 411 (2001) 521.

[21] S. Lawrence and G.L. Giles, Searching the World Wide Web, *Science* Vol. 280 (1998) pp. 98-100.

[22] S. Lawrence and G.L. Giles, Context and page analysis for improved Web search, *IEEE Internet Computing* Vol. 2 (1998) pp. 38-46.

[23] S. Lawrence, G.L. Giles and K. Bollacker, Digital libraries and autonomous citation indexing, *IEEE Computer* Vol. 32 (1999) pp. 67-71.

[24] M. Lesk, SIGIR'91: the more things change, the more they stay the same, *SIGIR Forum* Vol. 25 (1991) pp. 4-7.

[25] T.Y. Lin, Y.Y. Yao, and L.A. Zadeh (eds.), *Data Mining, Rough Sets and Granular Computing* (Heidelberg, Physica-Verlag, 2002).

[26] W. Marek and H. Rasiowa, Gradual approximating sets by means of equivalence relations, *Bulletin of Polish Academy of Sciences, Mathematics* Vol. 35 (1987) pp. 233-238.

[27] M.E. Maron and J.L. Kuhns, On relevance, probabilistic indexing and information retrieval, *Journal of the Association for Computing Machinery* Vol. 7 (1960) pp. 216-244.

[28] A. Odlyzko, The rapid evolution of scholarly communication, http://www.si.umich.edu/PEAK-2000/odlyzko.pdf

[29] Z. Pawlak, *Rough Sets, Theoretical Aspects of Reasoning about Data* (Dordrecht, Kluwer Academic Publishers, 1991).

[30] J.A. Pomykala, A remark on the paper by H. Rasiowa and W. Marek: "Gradual approximating sets by means of equivalence relations", *Bulletin of Polish Academy of Sciences, Mathematics* Vol. 36 (1989) pp. 509-512.

[31] S.E. Preece, Clustering as output option, *Proceedings of the American Society for Information Science* (1973) pp. 189-190.

[32] Y. Qiu and H. Frei, Concept based query expansion, *Proceedings of the Sixteenth ACM International Conference on Research and Development in Information Retrieval* (1993) pp. 160-169.

[33] V.V. Raghavan and H. Sever, On the resuse of past optimal queries, *Proceedings of SIGIR'95* (1995) pp. 344-350.

[34] E. Rasmussen, Clustering algorithms, in W. Frakes and R. Baeza-Yates (eds.) *Information Retrieval: Data Structures and Algorithms* (Englewood Cliffs, Prentice Hall, 1992) pp. 419-442.

[35] S.E. Robertson, M.E. Maron and W.S. Cooper, Probability of relevance: a unification of two competing models for document retrieval, *Information Technology: Research and Development* Vol. 1 (1982) pp. 1-21.

[36] S.E. Robertson and K. Sparck Jones, Relevance weighting of search terms, *Journal of the American Society for Information Science* Vol. 27 (1976) pp. 129-146.

[37] H. Sakai, K. Ohtake and S. Masuyama, A retrieval support system by suggesting terms to a user, *Proceedings 2001 International Conference on Chinese Language Computing* (2001) pp. 77-80.

[38] G. Salton, *Dynamic Information and Library Processing* (Englewood Cliffs, Prentice-Hall, 1975).

[39] G. Salton and M. McGill, *Introduction to Modern Information Retrieval* (New York, McGraw Hill, 1983).

[40] P. Schäuble and D. Knaus, The various roles of information structures, *Proceedings of the Sixteenth Annual Conference of the Gesellschaft für Klassifikation* (1993) pp. 282-290.

[41] R. Slowinski and D. Vanderpooten, A generalized definition of rough approximations based on similarity, *IEEE Transaction of Knowledge and Data Engineering* Vol. 12 (2000) pp. 331-336.

[42] K. Spark Jones, *Automatic Keyword Classification for Information Retrieval* (London, Butterworths, 1971).

[43] A. Tombros, R. Villa and C.J. van Rijsbergen, The effectiveness of query-specific hierarchic clustering in information retrieval, *Information Processing and Management* Vol. 38 (2002) pp. 559-582.

[44] E. Turban and J.E. Aronson, *Decision Support Systems and Intelligent Systems* (New Jersey, Prentice Hall, 2001).

[45] C.J. van Rijsbergen, *Information Retrieval* (London, Butterworths, 1979).

[46] J.R. Wen, J.Y. Nie and H.J. Zhang, Clustering user queries of a search engine, *Proceedings of the Tenth International World Wide Web Conference* (2002) pp. 162-168

[47] J.R. Wen, J.Y. Nie and H.J. Zhang, Query clustering using user logs, *ACM Transactions on Information Systems* Vol. 20 (2002) pp. 59-81.

[48] I. Willett, Query specific automatic document classification, *International Forum on Information and Documentation* Vol. 10 (1985) pp. 28-32.

[49] I. Willett, Recent trends in hierarchic document clustering: a critical review, *Information Processing and Management* Vol. 24 (1988) pp. 577-597.

[50] S.K.M. Wong and Y.Y. Yao, On modeling information retrieval with probabilistic inference, *ACM Transactions on Information Systems* Vol. 13 (1995) pp. 38-68.

[51] S.K.M. Wong, Y.Y. Yao and C.J. Butz, Granular information retrieval, in F. Crestani and G. Pasi (eds.) *Soft Computing in Information Retrieval: Techniques and Applications* (Heidelberg, Physica-Verlag, 2000) pp. 317-331.

[52] Y. Yang and J.O. Pedersen, A comparative study on feature selection in text categorization, *Proceedings the 14th International Conference on Machine Learning* (1997) pp. 412–420.

[53] Y.Y. Yao, Relational interpretations of neighborhood operators and rough set approximation operators, *Information Sciences* Vol. 111 (1998) pp. 239-259.

[54] Y.Y. Yao, Granular computing: basic issues and possible solutions, *Proceedings of the 5th Joint Conference on Information Sciences* Vol. I (2001) pp. 186-189.

[55] Y.Y. Yao, Information granulation and rough set approximation, *International Journal of Intelligent Systems* Vol. 16 (2001) pp. 87-104.

[56] Y.Y. Yao, Informaiton retrieval support systems, *Proceedings of the 2002 IEEE World Congress on Computational Intelligence* (2002) pp. 773-778.

[57] Y.Y. Yao, H.J. Hamilton and X. Wang, PagePrompter: an intelligent Web agent created using data mining techniques, *Proceedings of International Conference on Rough Sets and Current Trends in Computing*, LNAI 2475 (2002) pp. 506-513.

[58] Y.Y. Yao, K. Song and L.V. Saxton, Granular computing for the organization and retrieval of scientific XML documents, *Proceedings of the Sixth International Conference on Computer Science and Informatics* (2002) pp. 377-381.

[59] Y.Y. Yao and N. Zhong, Granular computing using information tables, in T.Y. Lin, Y.Y. Yao and L.A. Zadeh, (eds.) *Data Mining, Rough Sets and Granular Computing* (Heidelberg, Physica-Verlag, 2002) pp. 102-124.

[60] C.T. Yu, Adaptive document clustering, *Proceedings of SIGIR'85* (1985) pp 197-203.

[61] L.A. Zadeh, Towards a theory of fuzzy information granulation and its centrality in human reasoning and fuzzy logic, *Fuzzy Sets and Systems* Vol. 19 (1997) pp. 111-127.

[62] O. Zamir and O. Etzioni, Web document clustering: a feasibility demonstration, *Proceedings of SIGIR'98* (1998) pp. 46-54.

MIX
Papier aus verantwortungsvollen Quellen
Paper from responsible sources
FSC® C105338

If you have any concerns about our products,
you can contact us on
ProductSafety@springernature.com

In case Publisher is established outside the EU,
the EU authorized representative is:
Springer Nature Customer Service Center GmbH
Europaplatz 3, 69115 Heidelberg, Germany

Printed by Libri Plureos GmbH
in Hamburg, Germany